To Prof. Zyg,

Thanks very much for your warmly encouragement in environmental law study.

With best regards.

Qihao.

April 8, 2019.

Climate Change and Catastrophe Management in a Changing China

ELGAR STUDIES IN CLIMATE LAW

Series Editor: Jonathan Verschuuren, *Professor of International and European Environmental Law, Tilburg University, the Netherlands*

Climate change and responses to climate change are having an increasing effect on economic activities and peoples' lives. Across the globe, scholars are studying what these impacts mean or should mean for law. As a consequence, climate law is rapidly developing as a new and complex field of law, marked by the interactions of diverse areas such as migration law, human rights law, agricultural law, energy law, trade law, company law, tort law, insurance law, nature conservation law, marine law and so forth. This series publishes high quality scholarly works that make an essential and innovative contribution to the development of legal thinking on climate law and thus help us to understand how we can protect tomorrow's society against the consequences of climate change.

Titles in the series include:

The Concept of Climate Migration
Advocacy and its Prospects
Benoît Mayer

Climate Change and Catastrophe Management in a Changing China
Government, Insurance and Alternatives
Qihao He

Climate Change and Catastrophe Management in a Changing China

Government, Insurance and Alternatives

Qihao He

Associate Professor, College of Comparative Law, China University of Political Science and Law, Beijing, China

ELGAR STUDIES IN CLIMATE LAW

Edward Elgar
PUBLISHING

Cheltenham, UK • Northampton, MA, USA

Published by
Edward Elgar Publishing Limited
The Lypiatts
15 Lansdown Road
Cheltenham
Glos GL50 2JA
UK

Edward Elgar Publishing, Inc.
William Pratt House
9 Dewey Court
Northampton
Massachusetts 01060
USA

A catalogue record for this book
is available from the British Library

Library of Congress Control Number: 2018959280

This book is available electronically in the **Elgar**online
Law subject collection
DOI 10.4337/9781788111867

ISBN 978 1 78811 185 0 (cased)
ISBN 978 1 78811 186 7 (eBook)

Typeset by Servis Filmsetting Ltd, Stockport, Cheshire
Printed and bound in Great Britain by TJ International Ltd, Padstow

For Patricia McCoy,
who has supervised and supported me for many years

Contents

Figures

Tables

Foreword

Today, no thoughtful observer of the world of climate science any longer doubts that human-generated emissions of greenhouse gases (GHGs) are the primary cause of the controlling, threatening realities of climate change. Just the simple fact of sharply increasing atmospheric carbon concentrations over time—in the context of acute rapid weather shifts, super storms, and polar vortexes—makes clear that we are in an ascending spiral of geophysical, ecological, and economic risk.

China provides a deeply instructive focus for contemplating the years of increasing climate-induced troubles ahead. Since 2006 the People's Republic of China (PRC) has been the world's foremost emitter of greenhouse gases. Professor Qihao He's new book carries the reader through a thoughtful and expert examination of China's present environmental conditions and the economic buffering options that can begin to be implemented now to face and, however possible, mitigate the PRC's daunting climate realities. He convincingly illuminates the necessity for developing sophisticated private insurance structures as well as government structures for managing climate risk in China.

But Professor He's book is instructive far more broadly than the PRC. Given the increasing global convergence of economic as well as scientific interactions, the conclusions that Professor He reaches about climate risk-management within the world's vastest GHG emitter are likely to be relevant in some substantial part everywhere else on the planet as well.

Government risk-buffering must clearly be a major element in future climate risk management, but as Professor He shows, private market-based insurance structures are far more likely to provide expert, subtle, scientifically objective, broadly practicable, and effective risk anticipation and mitigation economic systems.

As past and recent precedent clearly demonstrates, government environmental management programs must live in a political world where the accuracy and societal importance of rapidly and dynamically changing physical facts and trends tend to get side-stepped, diluted, manipulated, denied, or ignored by official agencies due to suasions and dysfunctional interventions from the political world. The government of the United States, formerly a world leader in national environmental science and responsive law and policy, demonstrates the detrimental possibilities for governmental inconsistency, politically detouring its national climate policy into policy obscurantism, and sticking its multiple governmental heads into the sand.

Ultimately, geophysical realities will continue to assert themselves in unmistakable and unavoidable experience, and societally we must be on our best game merely to survive them. Professor He's book indicates how the design and structures of optimal potential climate risk management systems do exist, and implicitly predicts that eventually privately constituted risk-management insurance systems must be a dominant component of our societal climate-coping mechanisms.

Zygmunt J.B. Plater
Professor of Law
Coordinator of the Boston College Land & Environmental Law Program
Newton Centre, Massachusetts

Acknowledgements

At the moment of finalizing this book, I realized that one of the joys of completion is to reflect on the journey and remember all the people who have helped me along this academic road "less traveled by".

This book is the result of my studies at University of Connecticut School of Law (UCONN Law) and my work at China University of Political Science and Law (CUPL). My deep gratitude and appreciation go to Professor Patricia McCoy, who is now the Liberty Mutual Insurance Professor of Law in Boston College. She has been a guide both in writing this book and in life generally, and always makes me feel at home during my stay in Hartford and in Boston.

Special thanks to Professor Peter Siegelman (UCONN Law), Professor Tom Baker (Penn Law), Professor Michael Faure (Maastricht University) and Professor Zygmunt J.B. Plater (Boston College) whose comments and unrivalled understanding of insurance law/environmental law and economics have enlightened and sharpened my mind.

Special thanks to Professor Jun Wang, my supervisor from my undergraduate studies in China, and to Dr Ping Dai (who we call "teacher mother" in Chinese). They helped my wife Tian Tian (also their student) and I enormously with their enthusiasm.

Special thanks to Professor Xin Chen, who led me to the study of insurance law, and opened a new window to see the world. Special thanks to Yan Hong and her family, who helped me enormously with their hospitality. The wonderful time we had together to celebrate many Chinese holidays in West Hartford was precious and I will always have fond memories of that time.

I am sincerely grateful to the Ministry of Education Research Program for Young Scholars (2018年度教育部人文社会科学研究青年基金项目：气候巨灾保险的法律制度研究：以保险作为准政府机制的正当性为中心，项目批准号**18YJC820024**) and China University of Political Science and Law (中国政法大学青年教师学术创新团队支持计划资助18CXTD05) for financial support in finishing this book.

I owe the greatest debt to my family, my father Wuquan He, my mother Qiulan Li, my aunt Guilan He, my uncle-in-law Xiuxian Cheng, my sister Zhuoya He, my brothers Qilong He and Qimeng He, and my dear wife Tian Tian. Without their love, I could not have finished this book.

Some of the analysis in this book has appeared in prior journal and law review articles, including: Qihao He and Michael Faure, *Regulation by Catastrophe Insurance: A Comparative Study*, 24 Connecticut Insurance Law Journal 189 (2018); Qihao He, *Global Climate Governance and Disaster Risk Financing: China's Potential Roadmap in Transitional Reform*, China Legal Science, vol. 6: 28–49 (2018); Qihao He, *Regulation by Government-Sponsored Reinsurance in Catastrophe Management*, 23 Connecticut Insurance Law Journal 291 (2017); Qihao He, *Mitigation of Climate Change Risks and Regulation by Insurance: A Feasible Proposal for China*, 43 Boston College Environmental Affairs Law Review 319 (2016); Qihao He, *Climate Change, Catastrophe Risk and Government Stimulation of Insurance Market—A Study of Transitional China*, 17 International Finance Review 311 (The Political Economy of Chinese Finance) (2016); Qihao He, *Climate Change and Financial Instruments to Cover Disasters: What Role for Insurance?* in The Role of Law and Regulation in Sustaining Financial Markets 222 (Niels Philipsen et al. eds., 2015); and Qihao He and Ruohong Chen, *Securitization of Catastrophe Insurance Risk and Catastrophe Bonds: Experiences and Lessons to Learn*, 8 Frontiers of Laws in China 523 (2013). Thanks to Professor Michael Faure and Professor Ruohong Chen for allowing me to reprint our co-authored papers in this book.

Finally, my thanks go to Stephen Gutierrez, Caroline Kracunas, and Erin McVicar at Edward Elgar for their foresight, patience, assistance, and guidance in bringing this book to completion and ultimately to publication.

Qihao He
2019

Introduction*

I. BACKGROUND

My father is a peasant in Shandong Province in China. Every spring, as he planted his wheat fields, he became anxious because he knew from experience that drought, floods, and typhoons could threaten the harvest. Regularly, he complained to me that farming was captive to bad weather and he had no control. For my father and our family, each year's crop was just one weather disaster away from ruin.

At the time, neither my father nor I realized that the weather extremes he complained of were the early local warning signs of global climate change. Only later, during graduate studies, did I start to grasp the larger dimensions of the problem, both at home and abroad. When I learned that China is the world's leading emitter of greenhouse gases (GHGs), having surpassed US emissions in 2006, I decided to dedicate myself to the problem of global climate change.

The extraordinary growth of GHGs in China represents the single greatest challenge to global climate-change efforts in the coming decades. As Professor Alex Wang has noted, "[W]ithout a significant contribution from China, efforts to find a solution to global climate change are unlikely to succeed".[1]

Global climate change and the many weather-related catastrophes that have ensued have generated increasing losses. American International Group, Lloyd's of London, and other leading insurers all identify climate change as a major threat for global risk management.[2] Due to the combination of climate change and an increasing concentration

---

* The URLs cited throughout this text were last accessed on 5 December 2018.

[1] Alex Wang, *Climate Change Policy and Law in China*, in Oxford Handbook of International Climate Change Law 636 (Cinnamon P. Carlarne, Kevin R. Gray, and Richard Tarasofsky eds., 2016).

[2] AIG, *AIG's Policy and Programs on Environment and Climate Change* (2009), available at http://www.naic.org/documents/committees_ex_climate_survey_samp le_responses_AIG.pdf; Trevor Maynard, *Climate Change: Impacts on Insurers and How They Can Help with Adaptation and Mitigation*, 33 The Geneva Papers on Risk and Insurance—Issue and Practice 140 (2008).

of the world's population in vulnerable areas, it is likely that extreme weather-related disasters will become more frequent, more intense, and more costly in the coming years.[3] A 2006 report by the United Nations Framework Convention on Climate Change (UNFCCC) said that, by 2040, damage from climate change might be as high as US$1 trillion annually.[4] Moreover, the frequency of and losses from weather-related catastrophes has escalated since 1980 (Figure I.1 and Figure I.2).

Meanwhile, the impact of climate change on China has closely followed global trends. China has suffered some of the most severe natural catastrophes in the world, including floods, typhoons, and droughts. Direct economic losses caused by catastrophes in China are around $25 billion every year, and the number is considerably higher if indirect economic losses such as disaster relief are taken into consideration.[5] The Pearl River Delta is especially vulnerable to weather disasters, as it is a densely populated metropolitan area composed of Hong Kong, Guangzhou, and Shenzhen and is situated in one of the world's most disaster-prone regions. Floods and typhoons there put more people at risk than in any other metropolitan area in the world.[6]

II.   A BETTER WAY TO MANAGE CLIMATE CHANGE RISK IN CHINA

In the fight to combat global climate change, various legal and business mechanisms could be used to curb greenhouse gas emissions, distribute catastrophe risk and cover catastrophe losses. Among these mechanisms

---

[3]   Muthukumara Mani, Michael Keen and Paul K. Freeman, *Dealing with Increased Risk of Natural Disasters: Challenges and Options* (2003), available at http://www.imf.org/external/pubs/ft/wp/2003/wp03197.pdf.

[4]   See United Nations Environment Programme Finance Initiative (UNEP FI), Climate Change Working Group, *Adaptation and Vulnerability to Climate Change: The Role of the Finance Sector* 14 (2006), available at http://www.unepfi.org/fileadmin/documents/CEO_briefing_adaptation_vulnerability_2006.pdf. ("It seems very likely that the [sic] there will be a 'peak' year that will record costs over 1 trillion USD before 2040").

[5]   Department of Civil Affairs of People's Republic of China, Minzheng Shiye Fazhan Tongji Baogao (1990–2008), (2009), available at: http://news.xinhuanet.com/video/2009-05/24/content_11427652.htm, and http://wenku.baidu.com/view/788dd93067ec102de2bd89d2.html.

[6]   Swiss Re, *Mind the Risk: A Global Ranking of Cities under Threat from Natural Disasters* (2014), available at http://media.swissre.com/documents/Swiss_Re_Mind_the_risk.pdf.

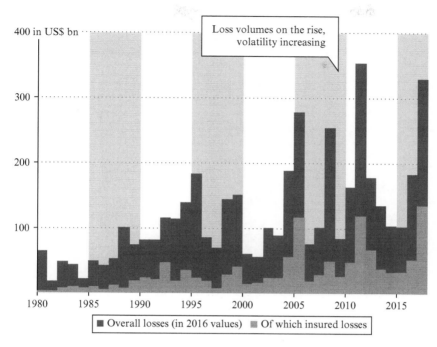

Source: Munich Re NatCatSERVICE, available at https://www.munichre.com/
topics-online/en/2018/01/2017-year-in-figures.

*Figure I.1   Losses from natural catastrophes, 1980–2017: volatility of loss
volumes is increasing*

are direct government intervention, private insurance, and insurance-
linked securities such as catastrophe (cat) bonds. There is still considerable
debate, however, about how to mitigate GHG emissions and compensate
catastrophe losses efficiently, effectively, and fairly—either singly or
through a combination of those mechanisms—especially for those countries and regions where many natural disasters occur, such as the United
States, China, and the European Union.

Policymakers' first instinct is to rely on government intervention and
controls. Much environmental regulation consists of prohibitions and
caps on GHG emissions. However, government regulation of climate
change has limitations, which are increasingly becoming apparent.
Effective regulation may be blocked by lack of consensus or political
opposition. For instance, President Trump's retreat from the Paris
Agreement backs off from the Obama-era promise to reduce America's

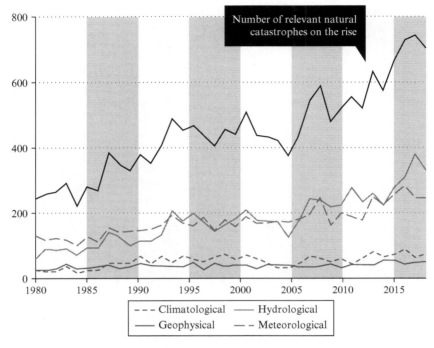

Source: Munich Re NatCatSERVICE, available at https://www.munichre.com/
topics-online/en/2018/01/2017-year-in-figures.

*Figure I.2   2017 joins the list of years with the highest number of natural
catastrophes*

emissions.[7] Furthermore, even when seemingly strict regulation is in
place, it may not work as planned. Government rules may have unintended consequences that exacerbate global warming. Companies, private citizens, and local officials often find ways to evade the rules, either
lawfully or unlawfully.

In view of these concerns, there is growing attention to how to augment
government regulation by harnessing market forces to address climate
change risks. First, regulators in certain countries, as part of imposing
emissions caps, have created markets in which heavy emitters can trade

---

[7]   See Michael D. Shear, *Trump Will Withdraw U.S. From Paris Climate
Agreement*, New York Times (June 1, 2017), available at https://www.nytimes.
com/2017/06/01/climate/trump-paris-climate-agreement.html.

emissions credits with low emitters. In that regard, in 2017 China released plans to start the world's largest emissions trading market.[8] Second, in the securities arena, many regulators require publicly listed companies to make periodic disclosures about their efforts to address climate change risks. Third, in the related corporate governance arena, institutional investors are increasingly pressing companies to recognize global warming risks and address them. For example, in the United States, stock insurers, including The Hartford and Travelers, face more pressure to confront climate change than mutual companies like American Family, Nationwide and State Farm because the stock companies are owned by sophisticated investors.[9] Last but not least, insurance has a growing role to play in compensating and mitigating global climate change risk, both through reimbursement of catastrophe losses and as a private regulator, by imposing loss mitigation requirements on insureds as a condition of coverage.

Increasingly, policymakers have come to realize that government alone cannot prevent or defray climate-related disaster risks. The role of insurance in managing global climate change has received a fair amount of attention in the United States and Europe. However, the topic has received less attention with respect to China, even though China is the world's largest greenhouse gas emitter and is taking a leading role in the implementation of the Paris climate change accord. Indeed, the Chinese mechanism for managing catastrophe risks, known as the "Whole-Nation System" (*Juguo tizhi*), entails neither private insurance nor social insurance, but rather a kind of emergency-driven disaster relief system. Therefore, this book examines the important but neglected question of how to enlist private- and public-sector insurance to compensate and mitigate climate change risk in China.

I contend that there is a better way to manage catastrophe risk in China through private insurance rather than directly through the government's Whole-Nation System. As a well-known risk management tool, private insurance could be an efficient financial instrument to cover catastrophe disasters. Compared to government-provided and subsidized compensation programs, private insurance offers advantages in the form of lower transaction costs, lower adverse selection, and greater efficiency as a result of competitive markets.

In addition, private insurance can act not only as a form of post-disaster

---

[8]   Keith Bradsher and Lisa Friedman, *China Unveils an Ambitious Plan to Curb Climate Change Emissions*, New York Times (December 19, 2017), available at https://www.nytimes.com/2017/12/19/climate/china-carbon-market-climate-change-emissions.html.

[9]   Joseph MacDougald and Peter Kochenburger, *Insurance and Climate Change*, 47 John Marshall Law Review 719, 730–745 (2013).

relief but also as a form of private regulation—a contractual device controlling and motivating behavior prior to the occurrence of losses.[10] When insurers underwrite catastrophe risk, they have strong incentives to mitigate risks and the ensuing losses in order to reduce their payouts. Regulating policyholders' behavior through risk-based pricing, contract design (e.g., deductibles and copayments), loss prevention, claims management, and refusals to insure, private insurance may function as a substitute for, or complement to, government regulation of catastrophe risks. In sum, insurance could not only compensate the victims of climate hazards but also reduce climate-change risks to begin with.

At the same time, promoting insurance to combat climate change risks faces a number of general challenges no matter what the country. Underwriting catastrophe insurance faces both supply-side and demand-side barriers. On the supply side, problems with insurability and capacity hamper the underwriting process. Private insurers often lack sufficient incentives or capacity to provide adequate, affordable catastrophe insurance or to mitigate climate change risk. Meanwhile, on the demand side, consumer demand for catastrophe insurance may also fall flat, due to the frequent high cost of the insurance, behavioral anomalies and repeated government bailouts. While these problems are not insuperable, they require careful thought and design to overcome.

On top of these general challenges, China faces additional hurdles, which are specific to the Chinese context, to making insurance a successful tool in fighting climate change. First, due to China's recent transition from a centralized planned economy to a market-based economy, the private property insurance and catastrophe insurance markets are still relatively new. This in turn depresses consumer demand for this insurance. Most citizens do not have homeowners' insurance (which is a major vehicle for weather catastrophe insurance in the United States). In addition, Chinese citizens are used, under the Whole-National System, to relying on the government, not private insurance, for catastrophe relief. Chinese consumers are even more reliant on government bailouts than consumers in developed market economies such as the United States because of China's recent experience with a state-controlled economy and the lack of a private insurance sector from the early 1950s through 1978.

This leads to a second hurdle faced by China, which is the propensity for government bailouts in weather disaster scenarios. Historically, the Chinese government has devoted great attention to preventing and

---

[10]   Omri Ben-Shahar and Kyle Logue, *The Perverse Effects of Subsidized Weather Insurance*, 68 Stanford Law Review 571 (2016).

distributing catastrophe losses. Today, the government still plays a major role in distributing catastrophe risk and compensating victims. Furthermore, under the authoritarian regime, natural disasters provide unique opportunities for the Chinese government to show its responsibility and accountability to the people. Therefore, the government has strong incentives to perform well and to increase people's loyalty through the Whole-Nation System.

Third, the capacity of the private insurance industry to supply weather catastrophe coverage is still limited. The total capital of China's property insurance companies is much lower than the total amount of losses caused by natural disasters.

Finally, the lack of weather catastrophe data makes it difficult for Chinese insurers to identify, quantify, and estimate the chance of disasters and to set premiums for catastrophe risks. Up to now the acquisition of such data has not been completed in China and remains to be accomplished.

Sophisticated techniques such as reinsurance and insurance-linked securities can help strengthen the capacity of the private insurance industry in China. However, both of these techniques are relatively new in China and lack sufficient backup resources. At present, China Re is the only state-owned reinsurance group in a monopoly position in China, occupying about 80 percent of the aggregate reinsurance market in China, with capital of $6.068 billion. Expanded entry by private global reinsurers could help redress this problem but this raises questions of foreign financial influence and power that may be of concern to the government. Although China's reinsurance market became open to foreign reinsurance companies after China's entry into the World Trade Organization, only a few reinsurance companies, such as Swiss Re and Munich Re, have established business operations in China, and they are only in the early stages of reinsuring risks.

In view of these challenges, support from the Chinese government will likely be necessary in order for insurance to serve as a meaningful tool in combating global climate change. The large scale of catastrophes means that the government will almost always have some role to play in their management. I argue, however, that that role should be indirect and regulatory, such as sponsoring government-backed reinsurance rather than providing direct compensation for losses, as is currently the case. This raises the question whether government participation will undermine the normal market discipline that private insurance exerts, in terms of proper pricing of premiums, deductibles and co-insurance, and the threat of cancellation if insured parties engage in moral hazard.

Compared to the United States and European Union, China has a distinctive history and political-economic configuration as a transition economy.

In order to fully develop a market-based catastrophe insurance system, which does not yet exist in China, a dynamic relationship between private sector insurance and the government is also crucial. Specifically, the government needs to take various measures to enhance the development of the private insurance industry. To help increase the capacity of the catastrophe insurance industry, the government also needs to boost the development of the reinsurance industry and insurance-linked securities market.

While there are no perfect answers, this book seeks to provide a roadmap to navigating these challenges and to tapping private and social insurance in the fight to address global warming and related disasters in China.

To be clear, extreme weather-related disasters, such as hurricanes, floods, typhoons, and snowstorms, are often classified as catastrophes.[11] Scholars and agencies have proposed many definitions of catastrophe risk.[12] In this book, catastrophe disasters are defined as natural events that occur infrequently but that cause significant human and financial losses.

---

[11]   The distinction between fundamental and particular risks is based on the discussion of hazard by Kulp. Fundamental risks are those caused by conditions more or less beyond the control of individuals and involve losses affecting a large proportion of the population. See C.A. Kulp, Casualty Insurance: An Analysis of Hazards, Policies, Insurers, and Rates 3–4 (1956).

[12]   For example, scholar Véronique Bruggeman defines such risk as a rapid onset, single-event disaster that causes a substantial amount of damage and/or that involves numerous victims. Erik Banks has expanded the definition from the traditional view of a single event that causes sudden changes to include instances of a gradual accumulation of many small incidents, perhaps precipitated by the same catalyst, leading to the same scale of damages/losses; such events may not actually be recognized as catastrophes until a long period of time has passed and many losses have accumulated. From the government perspective, for example, the Centre of Research on the Epidemiology of Disasters (CRED) treats catastrophe risk as "a situation or event, which overwhelms local capacity, necessitating a request to national or international levels for external assistance; an unforeseen and often sudden event that causes great damage, destruction and human suffering". According to Federal Emergency Management Agency (FEMA) of United States, an event where related federal costs reach or exceed $500 million is deemed a "catastrophe". See Véronique Bruggeman, Compensating Catastrophe Victims: a Comparative Law and Economics Approach 7 (2010); Erik Banks, Catastrophe Risk, Analysis and Management 5 (2005); The Em-Dat Glossary, The International Disaster Database (2012), available at http://www.emdat.be/glossary/9#term81; US Government Accountability Office, Experiences from Past Disasters Offer Insights for Effective Collaboration after Catastrophe Events 2 (2009).

III.   RESEARCH METHODOLOGY AND CONTRIBUTION

This book uses a law and economics methodology to analyse these issues. For instance, I develop and apply a supply-demand framework to analyse the feasibility of private insurance in distributing catastrophe risk in China. In addition, I will explore consumer anomalies in the areas of private and social catastrophe insurance based on theories of individual bounded rationality and cognitive biases.[13]

This book also uses comparative legal analysis to compare five middle- to upper-income countries—the United Kingdom, the United States, France, Japan, and Turkey—where catastrophic risks are more or less regulated by insurance. Based on this comparative analysis, I conclude that the countries where insurers are most successful in stimulating disaster risk reduction are also those where the insurers as private regulators complement and build upon effective public regulation.

The book's unique contribution lies in explaining how private sector insurance could be harnessed to better protect China from climate change risks. While other scholars have studied the science and incidence of climate change risk in China, no scholar to date has addressed the role of finance in paying for and reducing climate-change-induced disaster risk in the face of the unique challenges posed by China. This problem is of first-order magnitude because of China's size and its heightened exposure to climate change disasters. China faces three special challenges in providing compensation for such risk: (1) under the socialist system, the Chinese public became used to depending on the government to pay for disaster relief; (2) the high cost of continued government disaster relief is unsustainable; and (3) private sector insurance in China is still

---

[13]   Behavioral law and economics is based on cognitive psychology, which shows that human decision-making deviates from pure rational thinking. The task of behavioral law and economics is to explore the implications of actual (not hypothesized) human behavior for the law. See Russell B. Korobkin and Thomas S. Ulen, *Law and Behavioral Science: Removing the Rationality Assumption from Law and Economics*, 88 California Law Review 1051 (2000); Christine Jolls, Cass R. Sunstein and Richard Thaler, *A Behavioral Approach to Law and Economics*, 50 Stanford Law Review 1471 (1998); Joshua D. Wright and Douglas H. Ginsburg, *Behavioral Law and Economics: Its Origins, Fatal Flaws, and Implications for Liberty*, 106 Northwestern University Law Review 1033 (2012); Grant M. Hayden and Stephen E. Ellis, *Law and Economics After Behavioral Economics*, 55 University of Kansas Law Review 629 (2007); A.C. Dailey and P. Siegelman, *Predictions and Nudges: What Behavioral Economics Has to Offer the Humanities, and Vice-Versa*, 21 Yale Journal of Law & Humanities 341 (2013); Cass R. Sunstein, Behavioral Law and Economics (2000).

in its infancy. This book would be the first work to examine how a new approach to private sector and social insurance in China could compensate for catastrophe risks from climate change while reducing the magnitude of those risks. The book's comprehensive policy solution for China should inform and assist legislators and policymakers in developing new, market-based climate change policies for China and for other developing countries that face similar challenges.

IV.   ROADMAP OF THE BOOK

This book is broken into three main parts. The first part (Chapter 1) analyses the Chinese government's role as the nation's catastrophe risk manager and focuses on the state-run Whole-Nation System in China. Chapter 1 also explains why government regulation of climate catastrophe risks alone is less effective than when it is combined with regulation by private insurance.

The second part is the heart of this work (Chapters 2 through 6). Chapter 2 discusses the role that insurance plays in providing compensation as a form of post-disaster relief. Policyholders may substitute a small certain cost, the premium, for a large uncertain financial loss. However, due to the low-probability but high-consequence nature of catastrophe risk, catastrophe insurance faces inefficiencies both in terms of supply and demand. Risk-based pricing, mandatory multiyear insurance addressing the lack of demand and multiyear insurance enlarging insurers' supply deserve more attention as solutions to these inefficiencies.

Chapter 3 turns to the mitigation role of insurance. This chapter compares liability insurance and catastrophe insurance in mitigating climate-change risks that cause extreme weather disasters. Since liability insurance relies on the efficiency of tort-based climate-change litigation, which is still in its infancy, liability insurance is currently infeasible as a mechanism to regulate the emission of GHGs and thus help mitigate climate-change risks. First-party catastrophe insurance is a more feasible method since it mainly focuses on regulating the accumulation of value-at-risk. I conclude that compulsory catastrophe insurance could not only provide catastrophe disaster victims with financial protections, but also enhance mitigation of value-at-risk.

Chapter 4 further explores catastrophe insurance as a form of private regulation in managing catastrophe risks. This chapter identifies five regulatory techniques of catastrophe insurance that create incentives for optimal behavior by policyholders and examines how first-party insurance operates to regulate catastrophe risk in five middle- to high-income countries. This chapter suggests that regulation of catastrophe insurance may have a significant positive effect on catastrophe risk mitigation in China if the market for catastrophe insurance substantially expands.

Chapter 5 examines the important role of reinsurance—which essentially is insurance for insurers and spreads the underwriting risks of primary insurers—in increasing capacity and bolstering loss mitigation. With regulatory techniques such as loss-sensitive premiums, the duty of utmost good faith, risk management services and the indirect regulation of insured parties, reinsurance offers primary insurers incentives to underwrite appropriately and mitigate risk. I suggest that the government should adopt government-sponsored reinsurance to address catastrophe risks and thereby act as the last resort for primary insurers while at the same time regulating their underwriting behavior properly.

Chapter 6 discusses an alternative financial vehicle for managing catastrophe risk, consisting of insurance-linked securities. Insurance-linked securities connect the insurance industry to the capital markets and offer the potential to expand insurers' capacity. Although insurance-linked securities have not developed as quickly as predicted, policymakers should not overlook their advantages in solving problems of catastrophe insurance and covering disaster losses.

The final part of the book (Chapter 7) weaves the entire analysis into a plan for reform and proposes policy recommendations for China. The Chinese government's Whole-Nation System needs reform and a shift to an insurance-based model. To accomplish this shift, the Whole-Nation System (regulation by government) should be combined with catastrophe insurance (regulation by insurance) to capitalize on the strengths of the private market and public government oversight. Under such a hybrid system, the government should provide support to make catastrophe insurance by the private sector the key vehicle for the finance and loss mitigation of climate change risks. This chapter identifies solutions for solving the lack of supply and demand of catastrophe insurance, and also for increasing risk mitigation.

1. Climate change, catastrophe risk, and government stimulation of the insurance market—a study of transitional China

Insurance is something we tend to think about only after a disaster.[1]

I. INTRODUCTION

Due to climate-related extremes, growing populations in high-risk areas, and aging infrastructure but low levels of public and private investment in risk reduction measures, the world is more vulnerable to catastrophe disasters, and the losses are increasing significantly.[2] How to manage catastrophe risk efficiently and cover disaster losses fairly is still a universal dilemma.

"Famine happens every three years, epidemic happens every six years, and natural hazard happens every twelve years."[3] This old saying is a perfect description of natural disasters in China. Due to China's unusual size and regional diversity, as well as its distinctive history and current political-economic configuration of "socialism with Chinese characteristics,"[4] its approach to handling disaster challenges is in many

---

[1] Adam F. Scales, *Nation of Policyholders: Governmental and Market Failure in Flood Insurance*, 26 Mississippi College Law Review 3, 3 (2006).

[2] Erwann Michel-Kerjan, *Have We Entered an Ever-Growing Cycle on Government Disaster Relief*, Presentation to US Senate Committee on Small Business and Entrepreneurship (2013).

[3] Y.-M. Wei, J.-L. Jin and Q. Wang, *Impacts of Natural Disasters and Disasters Risk Management in China: The Case of China's Experience in Wenchuan Earthquake*, in Economic and Welfare Impacts of Disasters in East Asia and Policy Responses 641 (Y. Sawada and S. Oum eds., 2012).

[4] "Socialism with Chinese characteristics" is a grand but marvelously vague expression. It was first raised by reformist politician Deng Xiaoping in 1984, and it stretches the acceptable ideological framework to allow the country to pursue

ways unique.[5] The Chinese mechanism for managing catastrophe risks or challenges is known as the "Whole-Nation System" (*Juguo tizhi*), which generally refers to the government's effort to deploy and allocate the whole nation's resources to fulfill a specific difficult task within a limited time and thus promote the nation's interest.[6] The Whole-Nation System is not only a mechanism for coping with catastrophe disaster risks, but also a much larger political ideology for dealing with all kinds of landmark challenges. Besides catastrophe disaster relief, the Lunar Exploration Program and the Olympic Gold Medals Strategy are also famous examples of the Whole-Nation System.

In the field of natural disasters, the Whole-Nation System is a kind of emergency-driven management system and focuses on disaster emergency relief. To be clear, it is neither private insurance nor social insurance, but rather a government subsidy program. While effective in delivering emergency relief in the short term, it is not sustainable in terms of compensation in the long term due to certain government failures, which this chapter will describe. Comparing the Whole-Nation System with possible alternatives is illuminating and may provide important insights for its reform.

Private insurance is traditionally regarded as a major mechanism to cover losses caused by disasters. Even though underwriting catastrophe insurance is susceptible to market failures because of both supply-side and demand-side barriers,[7] it is still an attractive tool to comprehensively deal with catastrophe risk, especially compared with government intervention, due to its advantages of lower transaction costs, lower adverse selection, and greater efficiency as a result of competitive markets.[8]

In this chapter, I try to assess the Chinese government's responsibilities

---

policies that worked. See Ezra F.Vogel, Deng Xiaoping and the Transformation of China 465 (2011).

[5]   Elizabeth J. Perry, *Growing Pains: Challenges for a Rising China*, 143 Daedalus 5 (2014).

[6]   Peijun Shi and Xin Zhang, *Chinese Mechanism against Catastrophe Risk—the Experience of Great Sichuan Earthquake*, 28 Journal of Tsinghua University (Philosophy and Social Sciences) 96 (2013).

[7]   Qihao He, *Climate Change and Financial Instruments to Cover Disasters: What Role for Insurance?* in The Role of Law and Regulation in Sustaining Financial Markets 228 (Niels Philipsen et al. eds., 2015).

[8]   See Dwight Jaffe and Thomas Russell, *Catastrophe Insurance, Capital Markets, and Uninsured Risks*, 62 Journal of Risk and Insurance 225 (1997); Howard Kunreuther, *The Case for Comprehensive Disaster Insurance*, 11 Journal of Law and Economics 133 (1968); Howard Kunreuther, *Mitigating Disaster Losses through Insurance*, 12 Journal of Risk and Uncertainty 171 (1996); George L. Priest, *The Government, the Market, and the Problem of Catastrophe Loss*, 12 Journal of Risk and Uncertainty 219 (1996).

under the Whole-Nation System, highlighting what is unique or unusual (for better or worse) in efforts to resolve the universal dilemma of catastrophe risks. Furthermore, I will discuss the Chinese government's responsibilities for embracing and developing catastrophe insurance (although cognizant of some market failures), and then propose a catastrophe insurance market-enhancing framework that marries the merits of both private market and public government to address the universal dilemma of natural disasters.

II.   NATURAL DISASTERS AND THEIR IMPACTS IN CHINA

A.   Overview

China routinely suffers some of the most severe natural catastrophes in the world. Floods, droughts, earthquakes, typhoons, and landslides/mudslides are the five most frequently occurring types.[9] Floods hit the eastern part of China almost every year, and drought is perhaps the most severe natural disaster for agriculture, especially for North China.[10] Historical records show that from 206 BC to AD 1936 there were 1,037 floods and 1,035 droughts in China, averaging 0.967 floods or droughts per year.[11] Typhoons affect the coastal areas in the southeast, much as hurricanes do in the United States. Earthquakes occur in the western and northern areas. For example, the Great Sichuan Earthquake struck the southwest and caused 69,277 deaths and around $100 billion in losses in 2008.[12] Since the founding of the People's Republic of China (PRC) in 1949, a "hazard cycle," which describes the phenomenon that a catastrophic disaster occurs every three years on average, has clearly existed based on data calculations.[13] As early as the 1920s, Mallory described China bluntly as a "Land of Famine".[14]

---

[9]   Xian Xu and Jiawei Mo, *The Impact of Disaster Relief on Economic Growth: Evidence from China*, 38 The Geneva Papers on Risk and Insurance—Issues and Practice 495 (2013).
[10]   Walter H. Mallory, China: Land of Famine I–XVI (1926).
[11]   Y.T. Deng, History of Relief of Famines in China [Zhongguo Jiuhuang Shi] 38 (1937).
[12]   Hu Jintao, *Address on the National Earthquake Relief Summary Commendation Conference* [*Zai Quanguo Kangzhen Jiuzai Zongjie Biaozhang Dahui Shangde Jianghua*] (2008), available at http://wenku.baidu.com/view/90a1b9b465ce0508763213de.html.
[13]   A.G. Hu, Natural Hazards and Economic Development in China [Zhongguo Ziran Zaihai yu Jingji Fazhan] 5–10 (1997).
[14]   Walter H. Mallory, China: Land of Famine (1926).

B.   Climate Change Leads to More Catastrophes

Climate change is occurring on a significant scale, and its effects are being felt on all continents and across the oceans.[15] It is demonstrated that there is a clear link between climate change and many extreme weather-related catastrophes: "[A] changing climate leads to changes in the frequency, intensity, spatial extent, duration, and timing of extreme weather and climate events, and can result in unprecedented extreme weather and climate events".[16]

The impact of climate change for China has closely followed the global trend. In 2005, it was estimated that by 2020, the national average surface temperature in China could increase by 1.7°C; by 2030, 2.2°C; and by 2050, 2.8°C.[17] The affected areas of climate warming extend from south to north.

The adverse consequences of climate change will be severe for China. More droughts will occur, the drought-prone area will continue to expand, and droughts will grow more intense. Except for the increased rainfall in the western part of the northwest, the northern and southern part of the northeast could become permanently dry.[18] Meanwhile, floods, heavy rainfall, and landslides are likely to increase dramatically in the east of China.[19] The increasing frequency and intensity of catastrophe disasters

---

[15]   IPCC, *Climate Change 2014: Impacts, Adaptation, and Vulnerability* (2014), available at http://www.ipcc.ch/report/ar5/wg2/.

[16]   IPCC, *Summary for Policymakers* in Managing the Risks of Extreme Events and Disasters to Advance Climate Change Adaptation 3 (C.B. Field, V. Barros, T.F. Stocker, D. Qin, D.J. Dokken, K.L. Ebi, M.D. Mastrandrea, K.J. Mach, G.-K. Plattner, S.K. Allen, M. Tignor and P.M. Midgley eds.), A Special Report of Working Groups I and II of the Intergovernmental Panel on Climate Change (Cambridge, UK: Cambridge University Press; New York, NY, USA) (2012).

[17]   D.H. Qin, Y.H. Ding, J.L. Su, J.W. Ren, S.W. Wang, R.S. Wu, X.Q. Yang, S.M. Wang, S.Y. Liu, G.R. Dong, Q. Lu, Z.G. Huang, B.L. Du and Y. Luo, *Assessment of Climate and Environment Changes in China (I): Climate and Environment Changes in China and Their Projection*, 2 (Suppl. 1) Advances in Climate Change Research 1 (2006).

[18]   D.H. Qin, Y.H. Ding, J.L. Su, J.W. Ren, S.W. Wang, R.S. Wu, X.Q. Yang, S.M. Wang, S.Y. Liu, G.R. Dong, Q. Lu, Z.G. Huang, B.L. Du, and Y. Luo, *Assessment of Climate and Environment Changes in China (I): Climate and Environment Changes in China and Their Projection*, 2 (Suppl. 1) Advances in Climate Change Research 1 (2006).

[19]   Based on the regression analysis of natural disaster occurrence and average global temperature from 1980 to 2010, the frequency of epidemics, extreme temperature, floods, and storms was projected to increase by 506 times per year if the average global temperature increases by 1°C. See Pan XB et al., *Natural Disaster Occurrence and Average Global Temperature*, 4 Disaster Advances 61 (2011).

will no doubt aggravate the vulnerability of the socioeconomic development of China.[20]

C.   Losses Caused by Catastrophic Disasters

China has become a booming economy—the second biggest in the world in 2009, with a per capita GDP of $6,100 in 2012.[21] (All amounts of money are in US dollars unless designated with RMB.) China also has the largest population in the world, and most of it lives in South China and East China, both of which are vulnerable to floods, typhoons, landslides/mudslides, and earthquakes. For example, the Pearl River Delta, a densely populated metropolitan area comprising Hong Kong, Guangzhou, and Shenzhen, is situated in one of the world's most disaster-prone regions. Floods and typhoons there put more people at risk than in any other metropolitan area in the world.[22]

As a consequence, the economic losses and the population affected by catastrophes are increasing significantly and causing much greater socioeconomic impacts in China. Relative to the period 1900–2012, China experienced the highest frequency of natural disasters from 2003 to 2012, accounting for 37.6 percent of total occurrences from 1900 to 2012. That decade also accounted for 55.5 percent of the economic losses and 52.4 percent of the affected population from 1900 to 2012 (Table 1.1). Data for almost the same time period (2004–2010) show that major natural disasters are frequent in China and losses caused by catastrophes there are significant (Table 1.2).

III.   THE WHOLE-NATION SYSTEM AND HOW IT WORKS IN CHINA

Since 1949, the state has traditionally played the major role in covering risk and providing disaster relief in China.[23] After launching the project

---

[20]   Sha Chen, Zhongkui Luo and Xubin Pan, *Natural Disasters in China: 1900–2011*, 69 Natural Hazards 1597 (2013).

[21]   Justin Yifu Lin, *Demystifying the Chinese Economy*, 46 The Australian Economic Review 259 (2013).

[22]   Swiss Re, *Mind the Risk: Cities under Threat from Natural Disasters* (2013), available at http://media.swissre.com/documents/Swiss_Re_Mind_the_risk.pdf.

[23]   Zhou Yanli (Vice President of China Insurance Regulatory Commission), speech at International Catastrophe Insurance Fund Management Symposium (2008), available at http://insurance.hexun.com/2008/jzfx/index.html.

Table 1.1   *Natural disasters of 2003–2012 in China, relative to the period 1900–2012 (percent)*

	Occurrences	Deaths	Affected population	Direct economic loses
Drought	25.0	0.0	31.2	28.4
Earthquake	34.1	10.4	76.0	92.2
Flooding	53.0	0.1	35.6	38.8
Landslide	41.8	54.0	96.3	48.5
Local storm	39.0	18.6	19.8	68.7
Tropical cyclone	32.5	1.3	55.8	56.3
Average	37.6	14.1	52.4	55.5

Notes:
Cited from Xian Xu & Jiawei Mo, The Impact of Disaster Relief on Economic Growth: Evidence from China, 38 The Geneva Papers on Risk and Insurance—Issues and Practice 495–520 (2013).
Data include data up to 31 January 2012, 'Past Decade' roughly refers to the period from 2003 to 2012.

Source: EM-DAT: THE OFDA/CRED International Disaster Database, www.emdat. be—Université catholique de Louvain, Brussels (Belgium). Calculations and categorisation performed by the author.

of "reform and opening" (*gaige kaifang*) in 1978, the Chinese government gradually established disaster prevention, reduction, and relief mechanisms—collectively known as the "Whole-Nation System".[24]

In contrast with federal disaster policy in the United States, the Whole-Nation System is not the result of the failure of the private catastrophe insurance market but rather the child of China's history, economy, and socialist political system. Before considering what the content of the Whole-Nation System is and how it works, it is useful to examine the history of China's disaster policy.

A.   Historical Review of the Whole-Nation System: The Transformation of Disaster Risk Management since 1949

Natural disaster management has been a highly sensitive issue for thousands of years in China. Building dams to prevent floods and protect

---

[24]   Ming Zou and Yi Yuan, *China's Comprehensive Disaster Reduction*, 1 International Journal of Disaster Risk Science 24 (2010).

Table 1.2   Occurrences of major natural disasters in China, 2004–2010

Disasters	2004	2005	2006	2007	2008	2009	2010	Total	Loss (bn $US)
Earthquake	5	2	6	1	7	2	5	28	86.79
Flood	9	11	20	12	7	7	5	71	315.29
Storm	7	14	8	6	9	10	6	60	26.37
Drought	0	2	1	0	1	2	0	6	6.74
Extreme tempreture	1	1	0	0	2	0	0	4	21.10

Note: Cited from Xian Xu and Jiawei Mo, *The Impact of Disaster Relief on Economic Growth: Evidence from China*, 38 The Geneva Papers on Risk and Insurance—Issues and Practice 495–520 (2013).

Source: EM-DAT: The OFDA/CRED international Disaster Database, www.emdat. be—Université catholique de Louvain, Brussels (Belgium). Calculation and Categotations performed by the author.

agricultural production was routinely regarded as a major function of the centralized government in ancient times.[25] Disaster management was also stipulated by law and regulations. For example, "Records of Laws and Systems of Qing Dynasty" (*Da Qing Hui Dian Shi Li*, 1899) listed 12 articles identified as "Disaster Defense and Reduction Policies".[26]

Since the founding of the People's Republic of China (PRC) in 1949, disaster risk management policy has passed through two phases. The socialist government and a centrally planned economy were key features of the post-revolutionary order and had great influence on disaster policy in the first phase. The dawn of the Third Plenum of the Eleventh Party Congress in 1978 brought an entirely new set of socioeconomic reforms, including disaster risk management reform.

---

[25]   Tong Xing and Zhang Haibo, *Disaster Management Analysis Framework Based on Chinese Problem [Jiyu Zhongguo Wenti De Zaihai Guanli Fenxi Kuangjia]*, 1 China Social Science 132 (2010); Karl August Wittfogel, Oriental Despotism: A Comparative Study of Total Power, I–XIX (1957).

[26]   The articles included but were not limited to the following: food supply, river control and levee building, eradication of locusts, information dissemination, and so on. See H. Chen, *Disaster Defense and Reduction Policies in the Qing Dynasty [Qingdai fangzai Jianzai De Zhengce Yue Cuoshi]*, 3 Studies in Qing History [Qingshi Yanjiu] 41 (2004).

1. Phase I: A centrally planned state (1949–1978)

The period from 1949 to 1978 saw the introduction of centrally planned disaster policies, accompanied by the shutting down of private insurance markets. During this period, people had no opportunity to buy insurance to cover their catastrophe exposures. The centrally planned economic system, not the mere existence of market failures, was responsible for the disappearance of private insurance. According to the requirements of socialist planned governance, there is no need—at least in theory[27]—for business insurance because the government will bear all risks and cover individuals' exposures.

Meanwhile, the central government promulgated "Regulation on the Organization of National Disaster Relief Commission" (*Zhongyang Jiuzai Weiyuanhui Zuzhi Jianze*), which stipulated disaster policy as "self-rescue of victims, mutual support within social networks, and necessary assistance by government relief".[28] Self-rescue of victims means citizens were expected to take care of themselves in times of disaster. Mutual support within social networks meant people should help each other within the local community. Although the government purported to cover individuals' exposures, the government lacked the capacity to compensate disaster victims. China was one of the poorest nations at that time. Even in 1978, the per capita income was only $154, less than one-third of the average in Sub-Saharan African countries.[29] Worse still, political leaders prioritized the development of large, heavy, advanced industries as they started building the nation,[30] but paid less attention to natural disaster relief.

2. Phase II: Reform and transition (since 1978)

Along with the transformation from a centrally planned to a market-oriented economy (the so-called "reform and opening") starting in 1978, the disaster policy also underwent significant changes. The most significant feature of this period (since 1978) has been the dramatically increased government intervention and expansion of disaster relief.

---

[27] According to the view of Karl Marx, "from each according to his ability, to each according to his need," with the full development of socialism, there will be enough to satisfy everyone's needs. See Karl Marx, Critique of the Gotha Program (1875), available at https://www.marxists.org/archive/marx/works/1875/gotha/ch01.htm.

[28] Gazette of the State Council of the People's Republic of China, Vol. 40 (1957).

[29] Justin Yifu Lin, *Demystifying the Chinese Economy*, 46 The Australian Economic Review 259 (2013).

[30] Justin Yifu Lin, *Demystifying the Chinese Economy*, 46 The Australian Economic Review 259 (2013).

At present, relief and reconstruction after earthquakes, floods, typhoons, and other disasters is mostly financed by the government, along with social donations through the Red Cross and other charity NGOs. Unlike the actions in Phase I, the government has put the protection of people's lives and property higher on its agenda and has given a prominent position to natural disaster risk relief in its economic and social development plans.[31]

The comparison between the Tangshan Earthquake (1976) in Phase I and the Great Sichuan Earthquake (2008) in Phase II clearly shows the expansion of government relief and its changing priorities for disaster relief. Back in 1976, the Tangshan Earthquake, magnitude 7.5, killed 242,769 people.[32] Central and local governments put RMB 4.3 billion into aid for disaster relief. In an effort to save face and assert its independence, the Chinese government refused offers of disaster aid and assistance from foreign countries and international organizations.[33] In 2008, the Great Sichuan Earthquake, magnitude 8.0, caused 69,277 deaths.[34] In contrast, central and local governments put RMB 128.7 billion into disaster relief, a 30-fold increase.[35] Furthermore, the attitude to foreign assistance had been completely reversed. Only two days after the earthquake, the Chinese government formally requested support from the international community and in total received $500 million to help the families affected.[36]

The most convincing explanation for this change relates to China's miraculous success in economic growth since 1978, when China launched the project of "reform and opening". With GDP growing at an average

---

[31]   Office of National Commission for Disaster Reduction, P.R. China, *China's Natural Disaster Risk Management*, in Improving the Assessment of Disaster Risks to Strengthen Financial Resilience 121 (A Special Joint G20 Publication by the Government of Mexico and the World Bank, 2012).

[32]   USGS, *Historical Earthquakes: Tangshan China*, available at http://earthqu ake.usgs.gov/earthquakes/world/events/1976_07_27.php.

[33]   Wenying Tong, *Comparative Study on Catastrophe Disaster Relief Mobilization Mode in China: Case study between Tangshan Earthquake and Sichun Earthquake*, 5 Jianghai Academic Journal 94 (2010).

[34]   Hu Jintao, *Address on the National Earthquake Relief Summary Commendation Conference [Zai Quanguo Kangzhen Jiuzai Zongjie Biaozhang Dahui Shangde Jianghua]* (2008), available at http://wenku.baidu.com/view/90a1b9 b465ce0508763213de.html.

[35]   Wenying Tong, *Comparative Study on Catastrophe Disaster Relief Mobilization Mode in China: Case study between Tangshan Earthquake and Sichun Earthquake*, 5 Jianghai Academic Journal 94 (2010) (even considering inflation (5.92 percent per year), there is still a five-fold increase).

[36]   Jiuchang Wei, Dingtao Zhao and Dora Marinova, *Disaster Relief Drivers: China and the US in Comparative Perspective*, 11 China: An International Journal 93 (2013).

of 9.8 percent per year, and international trade growing by 16.6 percent annually over the past 33 years, China became an upper middle-income country, and more than 600 million people escaped poverty.[37] Per capita GDP in China reached $6,100 in 2012.[38] With China's success in raising personal incomes, and individuals' relative risk aversion increasing with wealth, protection against the loss of existing income likely emerged as an increasingly significantly social objective.[39] When the government began to obtain sufficient resources for disaster relief, disaster risk management became more important in the nation's policy agenda.

To date, China has established natural disaster risk management mechanisms featuring the role of government rather than that of the private insurance market.[40] These mechanisms are collectively named the Whole-Nation System (*Juguo tizhi*). It is more like a government subsidized disaster relief program than social insurance. This could be defined through the discussion of the content of the Whole-Nation System.

B.   The Content of the Whole-Nation System

1.   Origin

The phrase "Whole-Nation System" is borrowed from China's Soviet-style sports system.[41] The Whole-Nation System was formed and promoted along with Chinese participation in the Olympic Games. It has taken China from no gold medals to the top of the tally—in the 2008 Beijing Olympic Games, China won 51 gold medals while the United States won 36 and Russia won 23—quenching the thirst of many Chinese for national pride.[42] Olympic competition is one example of the

---

[37]   Barry Naughton, *China's Economy: Complacency, Crisis & the Challenge of Reform*, 143 Daedalus 14 (2014).

[38]   Justin Yifu Lin, *Demystifying the Chinese Economy*, 46 The Australian Economic Review 259 (2013).

[39]   See Kenneth Arrow, *Lecture 2: The Theory of Risk Aversion* in Aspects of the Theory of Risk-Bearing 28 (1965).

[40]   Office of National Commission for Disaster Reduction, P.R. China, *China's Natural Disaster Risk Management*, in Improving the Assessment of Disaster Risks to Strengthen Financial Resilience 121 (A Special Joint G20 Publication by the Government of Mexico and the World Bank, 2012).

[41]   Under this system, the government seeks talented young children from across the country, and the nation's sporting resources are concentrated on those who aim to become world champions. See Liang Xiao-long, *The Whole-nation System: The Successful Road of the Chinese Athletic Sports*, 26 Journal of Guangzhou Physical Education Institute 1 (2006).

[42]   Qingju Wang, *From "the Whole Nation System" to "Chinese Model": Theoretical Demand Space of Chinese Sports System a Summary of the Forum*

Whole-Nation System working effectively in achieving a challenging objective in a short time.

The 1998 Yangtze River flood and the 2003 Severe Acute Respiratory Syndrome (SARS) epidemic were regarded as watershed events for the development of the Whole-Nation System in the area of disaster response. In 2007, the "Emergency Response Law" was promulgated. This law comprehensively stipulates emergency response plans, institutions, mechanisms, and legal systems, and emphasizes the dominant role of government in emergency response to natural disasters.[43]

2.   Operating agencies

The Whole-Nation System operates under the unified leadership of the State Council (the Central People's Government). The central government is responsible for the coordination and organization of catastrophic disaster risk management.[44] At the national level, the system is headed by the National Committee for Disaster Reduction (NCDR), which consists of 33 disaster-related member agencies (Figure 1.1). A vice premier of the State Council serves as the director of the NCDR, and the Minister of Civil Affairs acts as its secretariat. A board of experts serves as consultants for the NCDR. For specific disasters, the corresponding ministries or bureaus are responsible for governance and technical affairs.[45] For example, the China Earthquake Administration takes charge of governance in the case of earthquakes, and the Ministry of Water Resources takes charge of governance in the case of floods and droughts. These coordinating bodies not only provide decision-making services for the NCDR on disaster response and relief but also implement its decisions.[46] At the local level,

---

"Sports System Transition and Chinese Sports in the Future", 34 Journal of Sports and Science 1 (2013).

[43]   Yongqing Wang, *Legislative Background and Overall Thinking on Emergency Response Law* (December 10, 2007), available at http://www.gdemo.gov.cn/yjpx/ztjz/200712/t20071210_37179.htm.

[44]   Office of National Commission for Disaster Reduction, P.R. China, *China's Natural Disaster Risk Management* in Improving the Assessment of Disaster Risks to Strengthen Financial Resilience 121 (A Special Joint G20 Publication by the Government of Mexico and the World Bank, 2012).

[45]   Y.-M. Wei, J.-L. Jin and Q. Wang, *Impacts of Natural Disasters and Disasters Risk Management in China: The Case of China's Experience in Wenchuan Earthquake* in Economic and Welfare Impacts of Disasters in East Asia and Policy Responses 641 (Y. Sawada and S. Oum eds., 2012).

[46]   Jiang Lingling et al., *People's Republic of China: Providing Emergency Response to Sichuan Earthquake*, Technical Assistance Consultant's Report to Ministry of Civil Affairs, P.R. China and Asian Development Bank (2008).

corresponding organizations in accordance with the national level are also established.

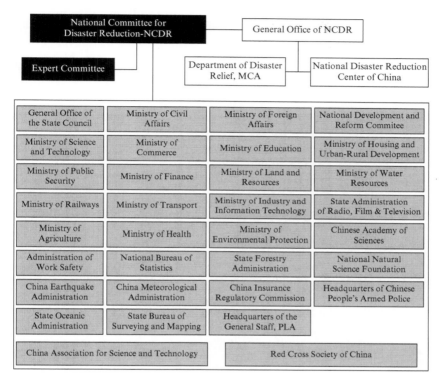

Source: Jiang Lingling et al., People's Republic of China: Providing Emergency Response to Sichuan Earthquake, Technical Assistance Consultant's Report to Ministry of Civil Affairs, P.R. China and Asian Development Bank (2008).

Figure 1.1   Disaster management organization structure

3.   Systematic arrangements

For a transition economy such as China, its legitimacy depends not only on its economic performance but also on its response and accountability to the people during disasters.[47] During the transition process, the Chinese

---

[47]   Y.A. Lazarev, A.S. Sobolev, I.V. Soboleva and B. Sokolov, *Trial by Fire: A Natural Disaster's Impact on Support for the Authorities in Rural Russia*, 66 World Politics 641, 642 (2014) ("The impact of natural disasters on support for

government has come to prioritize disaster relief in its agenda. Since the early 1980s, China has promulgated more than 30 laws and regulations concerning disaster prevention, reduction, and relief.[48] These laws and regulations cover different aspects of disaster risk management.[49] Besides laws and regulations, the Chinese government has also announced several national strategic plans relating to disaster risk management. As a recent example, in 2007, the Chinese government issued the National Plans for Comprehensive Disaster Reduction in the Eleventh Five-Year-Plan

---

authorities is conditional on governmental performance during and after the shock").

[48]   These laws and regulations include but are not limited to: Emergency Response Law of the People's Republic of China, Law of the People's Republic of China on Water and Soil Conservation, Law of the People's Republic of China on Protection against and Mitigation of Earthquake Disasters, Water Law of the People's Republic of China, Flood Control Law of the People's Republic of China, Law of the People's Republic of China on Desertification Prevention and Transformation, Meteorology Law of the People's Republic of China, Forestry Law of the People's Republic of China, Grassland Law of the People's Republic of China, Law of the People's Republic of China on the Prevention and Control of Water Pollution, Law of the People's Republic of China on the Prevention and Control of Pollution from Environmental Noise, Law of the People's Republic of China on the Prevention and Control of Environmental Pollution from Solid Waste, Marine Environment Protection Law of the People's Republic of China, Fire Control Law of the People's Republic of China, Drought Control Regulations of the People's Republic of China, Hydrology Regulations of the People's Republic of China, Flood Control Regulations of the People's Republic of China, Forest Fire Control Regulations of the People's Republic of China, Grassland Fire Control Regulations of the People's Republic of China, Regulations on Handling Major Animal Epidemic Emergencies, Regulations on the Prevention and Control of Forest Plant Diseases and Insect Pests, Regulations on the Prevention and Control of Geological Disasters, Regulations on the Handling of Destructive Earthquake Emergencies, Regulations on the Administration of Security of Reservoirs and Dams, and Regulations on the Administration of Weather Modification. See Information Office of the State Council of the People's Republic of China, *White Paper: China's Actions for Disaster Prevention and Reduction* (2009), available at http://mzzt.mca.gov.cn/article/fzjzgjhx/bps/200905/20090500030686.shtml.

[49]   Ming Zou and Yi Yuan, *China's Comprehensive Disaster Reduction*, 1 International Journal of Disaster Risk Science 24 (2010). For example, the Flood Control Law was promulgated in 1997 and revised in 2009, enacted to prevent and control flood, to take precautions against and alleviate calamities by flood and waterlogging, and to maintain the safety of people's lives and property. The Law of the People's Republic of China on Protecting against and Mitigating Earthquake Disasters came into force as of March 1, 1998, and was revised in 2009, intended to protect against and mitigate earthquake disasters and ensure the safety of people's lives and property.

Period.[50] Although it is difficult to evaluate how well these laws, regulations, and plans have been implemented, at least they reflect the government's concern for this issue. They stand for the government's willingness and efforts to prioritize disaster risk management on its agenda.

Under the Whole-Nation System, the government occupies the dominant position. In practice, government intervention can be classified into pre-disaster and post-disaster arrangements.

- Pre-disaster arrangements
  These include conducting natural disaster risk investigation and zoning, establishing a natural disaster monitoring system and an early warning system, pushing forward natural disaster prevention projects, establishing National Comprehensive Disaster Reduction Demonstration Communities, drawing people's attention to disaster prevention through designating May 12 as National Disaster Prevention and Reduction Day.[51]
- Post-disaster arrangements
  These include mobilizing national resources to deal with natural disasters, coordinating government and NGOs to implement disaster relief, organizing counterpart aid (*duikou zhiyuan*) to help disaster-affected areas, implementing reconstruction plans, and so on.[52]

Government fiscal support serves as the major capital source for disaster relief and post-disaster reconstruction.[53] Government disaster relief can be categorized into three layers under the Whole-Nation System.

- Emergency response
  This includes rescuing victims; providing medical treatment to injured people; providing food and shelter for victims; engaging in

---

[50]   Ming Zou and Yi Yuan, *China's Comprehensive Disaster Reduction*, 1 International Journal of Disaster Risk Science 24 (2010).

[51]   Office of National Commission for Disaster Reduction, P.R. China, *China's Natural Disaster Risk Management*, in Improving the Assessment of Disaster Risks to Strengthen Financial Resilience 121 (A Special Joint G20 Publication by the Government of Mexico and the World Bank, 2012).

[52]   K. Dalen, H. Flatø, L. Jing and Z. Huafeng, Recovering from the Wenchuan Earthquake. Living Conditions and Development in Disaster Areas 2008–2011, 39 (Oslo: FAFO, 2012).

[53]   Office of National Commission for Disaster Reduction, P.R. China, *China's Natural Disaster Risk Management*, in Improving the Assessment of Disaster Risks to Strengthen Financial Resilience 121 (A Special Joint G20 Publication by the Government of Mexico and the World Bank, 2012).

water purification, sanitation, quarantine, and epidemic prevention; and restoring transportation and other infrastructure for public interest.

- Direct payment to victims
  This includes a three-month temporary living subsidy to disaster-affected people, compensation to victim's families, compensation to the injured, orphans, the elderly, and the handicapped, and subsidies to help farm workers reconstruct their houses.
- Counterpart aid and reconstruction support
  These include rebuilding houses for victims, making long-term grants and loans to restore commercial structures and businesses, and aiding the agricultural sector.

In short, unified leadership by the government is the foremost principle of the Whole-Nation System.[54] It is a disaster-management system in which the government mobilizes, deploys, and allocates the nation's resources to cope with catastrophes, compensate victims, and conduct reconstruction, thus promoting the welfare of victims and the nation's interests. These interests extend beyond coping with a given disaster, including also promoting the government's image and fostering good relations between the various levels of government and the people.[55] In other words, it is not insurance; instead, it consists of fiscal cash outlays.

C.   How the Whole-Nation System Works in Practice—a Case Study of the 2008 Great Sichuan Earthquake

The 2008 Great Sichuan Earthquake showed how the Whole-Nation System worked. In 2008, an earthquake, magnitude 8.0, struck the Sichuan Province and caused 69,277 deaths.[56] The losses exceeded $100 billion.[57] The Whole-Nation System played an essential role in coping with this catastrophic earthquake. This section provides a case study

---

[54]   He Wang, Research on Catastrophe Risk Insurance Mechanisms 65(2013).

[55]   He Wang, Research on Catastrophe Risk Insurance Mechanisms 65(2013).

[56]   Hu Jintao, *Address on the National Earthquake Relief Summary Commendation Conference [Zai Quanguo Kangzhen Jiuzai Zongjie Biaozhang Dahui Shangde Jianghua]* (2008), available at http://wenku.baidu.com/view/90a1b9b465ce050876 3213de.html.

[57]   Hu Jintao, *Address on the National Earthquake Relief Summary Commendation Conference [Zai Quanguo Kangzhen Jiuzai Zongjie Biaozhang Dahui Shangde Jianghua]* (2008), available at http://wenku.baidu.com/view/90a1b9b465 ce0508763213de.html.

Source: Barry Naughton, China's Economy: Complacency, Crisis & the Challenge of Reform, 143 Daedalus 14 (2014).

Figure 1.2   *China's budget share in GDP, 1978–2012*

of the Whole-Nation System in operation following the Great Sichuan Earthquake.

1.   Undertaking disaster relief and compensating victims

Since the transition from a planned economy to a market economy in 1979, China has grown to be the second largest economy in the world, and its government has gained the unique ability to undertake disaster relief. After 1995, social resources available to the government, particularly for the central government, grew enormously due to both overall economic development and fiscal reform. Tax revenues as a share of GDP illustrate this dramatic change (Figure 1.2). Considering the rapid growth of the GDP itself, real budgetary revenues were almost 20 times in 2012 what they had been in 1995.[58] In 2012, Chinese government revenues at $1.86

---

[58]   Budget revenues increased from 10.8 percent of GDP in 1995 to 22.6 percent of GDP in 2012, and the annual GDP growth averaged 9.8 percent over the period.

trillion were about equal to the US federal government on-budget revenues, which are estimated by the Congressional Budget Office at $1.97 trillion (excluding Social Security).

Thanks to its economic power, the Chinese government has the ability to play a facilitating role in disaster relief. In the immediate aftermath of the Great Sichuan Earthquake, which occurred on May 12, 2008, the central government appropriated $83 million to victims that evening; within one week, the central government supplied more than $400 million in earthquake relief.[59] Within four months, the government had created an emergency disaster relief fund in the amount of about $11 billion,[60] while the losses from the earthquake were around $100 billion. On September 23, 2008, the State Council announced a "Notice on the State Council Overall Planning for Post-Great Sichuan Earthquake Restoration and Reconstruction" (hereinafter "the Plan"). According to the Plan, more than $157 billion was to be allocated for restoration work in the 51 disaster-affected counties in the provinces of Sichuan, Gansu, and Shaanxi.[61]

Direct payment to victims from government disaster relief funds is an important feature of the Whole-Nation System. Disaster relief funds were spent on victims mainly in the following ways. First, the funds were used to supply a three-month temporary living subsidy to earthquake-affected people. From May 20, this subsidy was about RMB 10 plus 0.5 kg of grain product every day to those affected people who had no residence and no income.[62] Second, the fund supplied compensation to victims' families. The number of fatalities was more than 70,000, and the government provided each victim's family with RMB 5,000.[63] Third, it compensated the injured, orphans, the elderly living

---

See Barry Naughton, *China's Economy: Complacency, Crisis & the Challenge of Reform*, 143 Daedalus 14 (2014).

[59] Xinhua News Agency, *Central Government Appropriated $400 Million to Earthquake Relief* (May 15, 2008), available at http://news.xinhuanet.com/newscenter/2008-05/15/content_8180172.htm.

[60] Peijun Shi and Xin Zhang, *Chinese Mechanism against Catastrophe Risk— the Experience of Great Sichuan Earthquake*, 28 Journal of Tsinghua University (Philosophy and Social Sciences) 96 (2013).

[61] The State Council (23 September, 2008), available at http://www.gov.cn/zwgk/2008-09/23/content_1103686.htm.

[62] Jiang Lingling et al., *People's Republic of China: Providing Emergency Response to Sichuan Earthquake*, Technical Assistance Consultant's Report to Ministry of Civil Affairs, P.R. China and Asian Development Bank (2008).

[63] Jiang Lingling et al., *People's Republic of China: Providing Emergency Response to Sichuan Earthquake*, Technical Assistance Consultant's Report to Ministry of Civil Affairs, P.R. China and Asian Development Bank (2008).

alone, and the handicapped affected by the earthquake (RMB 600 per month). The government launched a special mechanism to mobilize 375 hospitals from 20 provinces to treat more than 10,000 seriously injured victims, providing RMB 28,000 in medical subsidies for each injured person. For those with minor wounds, around 374,000 people, their medical treatment was free of charge.[64] Fourth, the government supplied subsidies to help farmers reconstruct their houses. In early June, the State Council decided that it would pay an average of RMB 10,000 per household to farmers whose houses collapsed or were severely damaged or who became homeless in the earthquake-affected region including Sichuan, Gansu, and Shaanxi provinces, for the purpose of reconstructing their houses.[65]

Three years after the earthquake, when the recovery and reconstruction period outlined in the Plan were over, the surveys of victims showed that most households' living conditions had indeed recovered to pre-earthquake levels or better.[66] For example, housing and employment goals, as the first two objectives of the Plan, had been largely fulfilled. Almost everyone in the earthquake-affected area lived in a permanent house, with only 0.6 percent of households still living in temporary houses or tents.[67] The employment rate was indeed relatively high, with the unemployment rate at only 2 percent; and household income also increased considerably.[68]

2. Mobilizing military power for emergency disaster relief

The Chinese People's Liberation Army (PLA) has a long history of involvement in disaster relief and constitutes an integral part of the Whole-Nation System. Due to the central government's mobilization and

---

[64]   Jiang Lingling et al., *People's Republic of China: Providing Emergency Response to Sichuan Earthquake*, Technical Assistance Consultant's Report to Ministry of Civil Affairs, P.R. China and Asian Development Bank (2008).

[65]   Jiang Lingling et al., *People's Republic of China: Providing Emergency Response to Sichuan Earthquake*, Technical Assistance Consultant's Report to Ministry of Civil Affairs, P.R. China and Asian Development Bank (2008).

[66]   K. Dalen, H. Flatø, L. Jing and Z. Huafeng, Recovering from the Wenchuan Earthquake. Living Conditions and Development in Disaster Areas 2008–2011, 39 (Oslo: FAFO, 2012).

[67]   K. Dalen, H. Flatø, L. Jing, and Z. Huafeng, Recovering from the Wenchuan Earthquake. Living Conditions and Development in Disaster Areas 2008–2011, 39 (Oslo: FAFO, 2012).

[68]   K. Dalen, H. Flatø, L. Jing, and Z. Huafeng, Recovering from the Wenchuan Earthquake. Living Conditions and Development in Disaster Areas 2008–2011, 39 (Oslo: FAFO, 2012).

the PLA's command-and-control structure, the PLA made significant contributions to the operation of the Whole-Nation System, especially in the emergency response to disasters.

In recent years, the mounting frequency of and losses from natural catastrophes in China has placed increasing demands on the military to be deployed to domestic disaster relief under its powers to conduct "military operations other than war".[69] China, as a developing country, lacks sufficient civilian emergency management capacity compared with Western countries, and thus the military has been a powerful, energetic, and integral part of the Whole-Nation System. In 2005, the State Council and the Central Military Commission jointly promulgated the "Regulation on the Army's Participation in Disaster Rescue", which designated the PLA as the "shock force" in national responses to catastrophic disasters.[70]

The PLA contributes significantly to the efficiency of the Whole-Nation System in disaster emergency response and relief. First, the military responded with unprecedented speed to the earthquake and made the best use of the golden hours of disaster rescue to save as many as lives as possible.[71] Only 13 minutes after the earthquake, the PLA activated the military plan for handling emergency incidents. Within 10 hours, 12,000 PLA and People's Armed Police (PAP) soldiers arrived and undertook earthquake rescue; and on the next day, another 11,420 troops arrived by air transportation alone.[72] Second, the civilian government made use of the military's vertical command-and-control structure to improve relief efficiency under conditions of catastrophic losses. Within the PLA, a three-tiered command system was quickly set up to oversee the military's relief operation, and these ad hoc institutions were also subject to the leadership of the civilian government at the corresponding level.[73] Third, the military undertook wide-ranging relief activities, counting on its huge numbers of troops. The PLA is the largest army in the world, and it can

---

[69]   William Banks, *Who's in Charge: The Role of the Military in Disaster Response*, 26 Mississippi College Law Review 75 (2007).

[70]   Ministry of Defense of PRC, (2005) available at http://www.mod.gov.cn/policy/2011-09/26/content_4300880.htm.

[71]   Jian Zhang, *The Military and Disaster Relief in China: Trends, Drivers and Implications*, in Disaster Relief in the Asia Pacific: Agency and Resilience 79 (Minako Sakai et al. eds., 2014).

[72]   W. Chen, *Zhijing! Xinshiqi Zuikeai de Ren [Pay Tribute to the Most Beloved People in the New Period]*, 8 Dangshi Yanjiu [Journal of Party History] 1 (2008).

[73]   Jian Zhang, *The Military and Disaster Relief in China: Trends, Drivers and Implications*, in Disaster Relief in the Asia Pacific: Agency and Resilience 79–80 (Minako Sakai et al. eds., 2014).

Table 1.3   *Numbers of soldiers that have participated in major disaster relief and Olympics security since 1998*

Year	Event	PLA and PAP TROOPS	Reserve and Militia
1998	Major flooding of the Yangtze, Songhua, and Nen Rivers	300,000	5,000,000
2002	Flooding in Shanxi, Fujjian, and 19 other provinces	20,000	170,000
2003	Flooding of the Huai River in Jiangxi, Hunan, and Shanxi provinces	48,000	410,000
2008	Snow and ice storms in 21 provinces	224,000	1,036,000
2008	Earthquake in Wenquan, Sichuan	146,000	75,000
2008	Security for the Olympics	131,000	na
2010	Earthquake in Yushu, Sichuan	16,000	na
2010	Mudslides in Zhouqu, Gansu	7,600	na

Source: Taylor Fravel, *Economic Growth, Regime Insecurity, and Military Strategy: Explaining the Rise of Noncombat Operations in China*, 7 Asian Security 177–200 (2011).

quickly mobilize a large number of soldiers to handle disasters, as it has throughout the PRC's history (Table 1.3). As many as 146,000 troops were deployed, and they evacuated around 1.4 million people, provided medical treatment to 1.36 million injured people, and rescued 3,338 people.[74] Furthermore, they also restored road transportation; provided food and shelter for victims; and engaged in water purification, sanitation, quarantine, and epidemic prevention, and the like.[75] Even though the majority of PLA troops withdrew within three months of the earthquake, some engineering units stayed for another three months to assist with post-earthquake reconstruction.[76]

---

[74]   Jian Zhang, *The Military and Disaster Relief in China: Trends, Drivers and Implications*, in Disaster Relief in the Asia Pacific: Agency and Resilience 80 (Minako Sakai et al. eds., 2014).

[75]   Jian Zhang, *The Military and Disaster Relief in China: Trends, Drivers and Implications*, in Disaster Relief in the Asia Pacific: Agency and Resilience 80 (Minako Sakai et al. eds., 2014).

[76]   Jian Zhang, *The Military and Disaster Relief in China: Trends, Drivers and Implications*, in Disaster Relief in the Asia Pacific: Agency and Resilience 80 (Minako Sakai et al. eds., 2014).

3.   Counterpart aid (*Duikou Zhiyuan*): National "pooling" of catastrophe risk among provincial and local governments

Pooling is a fundamental mechanism in both public risk management and private insurance. Its basic principle lies in combining and spreading a sufficient number of exposures across as large a group as possible.[77] Counterpart aid (*duikou zhiyuan*) is a mechanism that, under the central government's organization, requires some provinces that have stronger economic power to assist and support the reconstruction of disaster-affected areas.[78] This aid is generally conducted on a one-to-one basis, under the principle of "one province helps one significantly affected county".[79] The match criterion set up by the central government is that the richest donor provinces will contribute to the hardest-hit victim areas, while the less wealthy provinces will be asked to do less, and the least wealthy provinces will not be assigned any victim areas. According to the Plan, 18 heavily affected counties in Sichuan Province were supported by 18 other provinces or municipalities. For example, Guangdong Province—the richest province measured by GDP—was responsible for the reconstruction of Wenchuan County, which was the epicenter of the earthquake and suffered the most severe losses. Shandong Province—the second richest province in 2008—was responsible for and supported the reconstruction of BeiChuan County, which was the neighbor of Wenchuan County and was also heavily hit by the earthquake. Meanwhile, Gansu Province, Guizhou Province, and other less wealthy provinces had no responsibility for counterpart aid.

Under the framework of counterpart aid, supporting provinces are required to donate 1 percent of their fiscal revenue from the preceding year to supported counties for reconstruction in the next three

---

[77]   David A. Moss, When All Else Fails: Government as the Ultimate Risk Manager 292–296 (2004); E.J. Vaughan and T.M. Vaughan, Fundamentals of Risk and Insurance 34–44 (2007).

[78]   Y. Lan, *A Study on the Counterpart Aid: A Cooperation Pattern between Local Governments with Chinese Characteristics [Duikou Zhiyuan: Zhongguo Tese De Difang Zhengfu Jian Hezuo Moshi Yanjiu]*, Thesis of the College of Political and Law of Northwest Normal University [XiBei Shifan Daxue Shuoshi Xuewei Lunwen] (2011); L. Zhao and Y.J. Jiang, *Analysis of Local Government Coordinated Assistance Modes [Difang Zhengfu Duikou Zhiyuan Moshi Fenxi]*, 2 Journal of ChengDu University (Social Science Edition) [ChengDu Daxue Xuebao (Sheke Ban)] 4, 25 (2009).

[79]   Y.-M. Wei, J.-L. Jin and Q. Wang, *Impacts of Natural Disasters and Disasters Risk Management in China: The Case of China's Experience in Wenchuan Earthquake* in Economic and Welfare Impacts of Disasters in East Asia and Policy Responses 641 (Y. Sawada and S. Oum eds., 2012).

Table 1.4   *Performance of some supporting provinces in counterpart aid in response to the 2008 Great Sichuan Earthquake (RMB billions)*

Supporting provinces or municipalities	Fiscal revenue (2007)	Required counterpart aid (1%)	Supported counties	Actual counterpart aid (2008)
Guangdong	278.526	2.785	Wenchuan	4.162
Jiangsu	223.666	2.236	Mianzhu	4.363
Zhejiang	164.949	1.649	Qingchuan	3.499
Shanghai	207.448	2.074	Dujiangyan	6.198
Beijing	164.964	1.649	Shifang	5.306
Liaoning	108.199	1.081	Anxian	2.6
Fujian	70.03	0.703	Pengzhou	3.318
Anhui	54.347	0.534	Songpan	1.889

Source: Zhongdong Hua, *The Effects Analysis of Counterpart Support to the Equalization of Basic Public Services: Taking Earthquake-Stricken Areas of Sichuan as a Case [Duikou Zhiyuan Cujin Jiben Gonggong Fuwu Zhundenghua Xiaoying Fenxi]*, 5 Xi'an Caijing Xueyuan Xuebao [Journal of Xi'an University of Finance and Economics] 75–81 (2010).

years.[80] One percent is the minimum requirement of the central government, but the supporting province may increase the donation at its discretion. In fact, some provinces gave more than a 1 percent share (Table 1.4). Under the authoritarian regime, political benefit is an important incentive for the supporting provinces' governors to increase the counterpart aid capital.[81] Under China's authoritarian regime, the political selection of officers depends on a competitive political tournament. Therefore, the governors of provinces have incentives to perform better in counterpart aid to win support from/attract the attention of the central government.

In short, counterpart aid can be regarded as a special "pooling" of catastrophe risk because it spreads specific natural disaster risk across selected provinces and improves social welfare. This mechanism differs from a central fund, which would require every province to give

---

[80]   Y.-M. Wei, J.-L. Jin and Q. Wang, *Impacts of Natural Disasters and Disasters Risk Management in China: The Case of China's Experience in Wenchuan Earthquake* in Economic and Welfare Impacts of Disasters in East Asia and Policy Responses 641 (Y. Sawada and S. Oum eds., 2012).

[81]   See Lian Zhou, *Administrative Subcontract [Xizheng Fabaozhi]*, 6 Society [Shehui] 1 (2014).

1 percent to a fund that could then be used across victimized areas. Counterpart aid linking donor and recipient areas in this way seems economically pointless, since it only imposes unneeded constraints on the flow of funds. However, due to China's unusual size and regional diversity, especially the huge economic imbalance among different provinces—for example, the GDP of the richest province in 2014 is 77 times that of the poorest[82]—requiring the richer provinces to contribute more in disaster relief emphasizes the concern for equality rather than economic efficiency.

4.   People relying on and trusting in the government

Given the history of a planned economy, people in China still have a strong reliance on government, especially in the aftermath of a catastrophe. According to a nationwide survey in China that covered 856 cities and counties in 2009, when asked, "Who should take the major responsibility to undertake the burden of disaster losses and pay the bill?" nearly 70 percent of respondents indicated that government should take the major responsibility and cover disaster losses, while only 6.6 percent believed community or individual families should be responsible.[83] Under the leadership of the government, the Whole-Nation System accords with preferences for relying on government to deal with catastrophes.

A lack of public trust would undermine the credibility and stability of the Whole-Nation System, and lead to inefficiency, unrest, or even failure.[84] Only if the government enjoys high levels of trust can it gain space and freedom to act and have the opportunity to prove that its actions are in the interests of the people.[85]

Under the authoritarian Chinese regime, natural disasters provide unique opportunities for government to show its responsibility and

---

[82]   China GDP Ranking among Provinces in 2014, available at http://www.mnw.cn/news/cj/715060.html.

[83]   Ming Wang, Chuan Liao, Saini Yang, Weiting Zhao, Min Liu and Peijun Shi, *Are People Willing to Buy Natural Disaster Insurance in China? Risk Awareness, Insurance Acceptance, and Willingness to Pay*, 32 Risk Analysis 1717 (2012).

[84]   K. Dalen, H. Flatø, L. Jing and Z. Huafeng, Recovering from the Wenchuan Earthquake. Living Conditions and Development in Disaster Areas 2008–2011, 39 (Oslo: FAFO, 2012).

[85]   Jan Delhey and Kenneth Newton, *Social trust: global pattern or Nordic exceptionalism?* Discussion paper for Wissenschaftszentrum Berlin für Sozialforschung (WZB) (2004).

accountability to the people.[86] Therefore, the government has strong incentives to perform well and increase people's loyalty through the Whole-Nation System. In addition, it is easier for China to deliver disaster relief efficiently because, unlike the United States with its federal system, China is a unitary state in which the central government has absolute impact on the performance of the local governments.[87] During disaster relief, local governments are not only supported but also supervised by the central government. This system has facilitated the operation of the Whole-Nation System, especially counterpart aid.

5. A short conclusion

Under the Whole-Nation System, China's government has unique resources at its disposal to manage the disaster response and recovery processes. It coordinates multiple levels of government and establishes national pools. It also mobilizes the military for emergency response to disasters. Though by no means a perfect risk manager, the Chinese government does take up the vacuum, as Nobel economist Kenneth Arrow suggested in another context, to "undertake insurance in those cases where [a private market for insurance], for whatever reason, has failed to emerge".[88]

D. The Problems with the Whole-Nation System

Now we are ready to ask whether the Whole-Nation System can possibly do any better than private insurance, which has already demonstrated significant market failures in covering catastrophe risk. As will become clear, the answer is a resounding "yes" in theory but a tantalizing "no" in reality. In fact, the government is by no means a perfect risk manager. Just as even the best market systems confront "market failures", government solutions confront many obstacles and problems, collectively known as "government failures". The Whole-Nation System is no exception.

According to surveys conducted after the 2008 Great Sichuan

---

[86]   Y.A. Lazarev, A.S. Sobolev, I.V. Soboleva and B. Sokolov, *Trial by Fire: A Natural Disaster's Impact on Support for the Authorities in Rural Russia*, 66 World Politics 641 (2014) (The legitimacy of government depends on not only its economic performance but also its response and accountability to the people. The impact of natural disasters on support for authorities is conditional on governmental performance during and after the event).

[87]   Lian Zhou, *Administrative Subcontract [Xizheng Fabaozhi]*, 6 Society [Shehui] 1 (2014).

[88]   Kenneth Arrow, *Uncertainty and the Welfare Economics of Medical Care*, 53 The American Economic Review 941(1963).

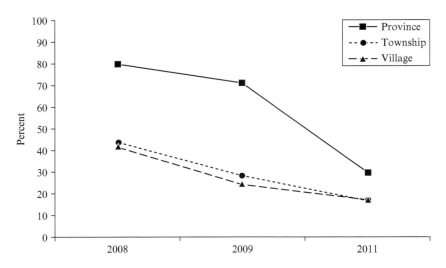

Source: K. Dalen, H. Flatø, L. Jing and Z. Huafeng, Recovering from the Wenchuan Earthquake. Living Conditions and Development in Disaster Areas 2008–2011, 175 (Oslo: FAFO, 2012).

Figure 1.3   Declining satisfaction with government relief after the 2008 Great Sichuan Earthquake

Earthquake, the degree of satisfaction with all levels of government (except the central government) had declined since the quake (Figure 1.3). The criticism of government relief grew over time.[89]

Added to the above discussion of how the Whole-Nation System worked during the 2008 Great Sichuan Earthquake must be an examination of some of its many problems, especially over the long run.

- The perverse incentives for rent-seeking and corruption
- Samaritan's Dilemma reducing people's incentives to invest in protection and mitigation measures
- The regressive effects of counterpart aid
- The lack of risk financing under the Whole-Nation System
- The burden on public budgets and possible hindrance to economic growth
- The overuse of the military's resources for nondefense purposes

---

[89]   K. Dalen, H. Flatø, L. Jing and Z. Huafeng, Recovering from the Wenchuan Earthquake. Living Conditions and Development in Disaster Areas 2008–2011, 175 (Oslo: FAFO, 2012).

1. Corruption problems

Before the economic reform and opening, corruption tended to be visible and easy to prevent because officials normally had only a single income source under the planned economy.[90] However, corruption problems became more serious and rampant when China decided to transit from a planned to a market economy. While the government remains as powerful as it has always been, the market-oriented reforms create substantially greater rent-seeking opportunities.[91] As a result, officials at every level have found many sources of income beyond a single government salary.[92] Officials who implement the Whole-Nation System are not easily immunized from corruption problems. In fact, increasing government aid under the Whole-Nation System may create more corruption opportunities than before.

There were several different forms of corruption among local officials in the wake of the 2008 Great Sichuan Earthquake. One form was the embezzlement or misuse of relief funds. The central government sent funds to various agencies in Sichuan Province after the earthquake. Although that money was intended for relief, significant amounts ended up paying for government banquets and officials' bonuses.[93] Another example of the embezzlement of relief funds was the case of a secondary school built in 2010 with quake relief funds, which was later torn down to make way for luxury houses.[94] A second form of corruption is the misappropriation

---

[90] Justin Yifu Lin, Demystifying the Chinese Economy 19 (2012).

[91] China's economic transition process is known as the "dual-track system" which differs from "shock therapy". The dual-track system refers to the process of moving from a single track, planned economic system to a combination system and finally to the single track of a market system. Under the dual-track system, market prices of goods during the transition are higher than the planned prices because planned prices are artificially suppressed by the government. Furthermore, government regulates market access to placate "legacy" suppliers. These market distortions too often lead to rent-seeking and corruption. See Justin Yifu Lin, Demystifying the Chinese Economy 194–198 (2012).

[92] For example, officials have the power to grant permits for a variety of businesses such as land acquisitions and construction. They may ask for direct payments but also shares in the company, property at below-market price, lavish dinners, and so on. Such practices are so widespread, and so many officials and their family members are involved, that those corruption problems are extremely difficult to solve. See Ezra F. Vogel, Deng Xiaoping and the Transformation of China 712 (2011).

[93] He Wang, Research on Catastrophe Risk Insurance Mechanisms 92–93 (2013).

[94] He Wang, Research on Catastrophe Risk Insurance Mechanisms 92–93 (2013).

or misuse of disaster goods. In Deyang, a city in Sichuan affected by the earthquake, an officer stole 10 containers of earthquake relief goods rather than send them to homeless residents.[95] Furthermore, some tents marked "disaster only" appeared on some officers' upscale patios, which had been barely touched by the quake.[96] A third form of corruption occurred when officials secretly granted reconstruction projects to their friends and family members without due process or transparency. According to the earthquake relief audit report of the National Audit Office of the PRC, there were 146 of these cases in Sichuan in 2008, totaling $220 million.[97]

Unfortunately, corruption problems were much worse in the long run under the Whole-Nation System than short-run emergency relief. The relief and reconstruction funds from the central government were allocated to local governments, and local governments were requested to take responsibility for implementation.[98] In the early days after the disaster, almost all government relief programs were under the spotlight of the whole nation, not only the central government but also the media. Public response to disasters is in proportion to media coverage. As is common throughout the world, media coverage surges upward in the immediate aftermath of a disaster, throwing a bright spotlight on the victims, and then attention quickly dissipates.[99] Under these circumstances, it is difficult for local officials to divert disaster aid.

However, as time passed and the media spotlight shifted, national attention was diverted. As a result, although the total accumulated capital for disaster relief was high, the central government leaders' focus on the disaster was short-lived.[100] This short time horizon weakened public and

---

[95]   *China Earthquake Brings Corruption, Relief Suspicion*, USA Today (May 29, 2008), available at http://usatoday30.usatoday.com/news/world/2008-05-29-china-corruption_N.htm.

[96]   *China Earthquake Brings Corruption, Relief Suspicion*, USA Today (May 5, 2008), available at http://usatoday30.usatoday.com/news/world/2008-05-29-china-corruption_N.htm.

[97]   He Wang, Research on Catastrophe Risk Insurance Mechanisms 92 (2013).

[98]   Jiang Lingling et al., *People's Republic of China: Providing Emergency Response to Sichuan Earthquake*, Technical Assistance Consultant's Report to Ministry of Civil Affairs, P.R. China and Asian Development Bank (2008).

[99]   David A. Moss, *The Peculiar Politics of American Disaster Policy: How Television has Changed Federal Relief* in The Irrational Economist 151 (Erwann Michel-Kerjan and Paul Slovic eds., 2010).

[100]   David A. Moss, *The Peculiar Politics of American Disaster Policy: How Television has Changed Federal Relief* in The Irrational Economist 151 (Erwann Michel-Kerjan and Paul Slovic eds.2010). For example, then-premier Wen Jiabao arrived at the affected areas to command disaster relief just eight hours after the 2008 Sichuan Earthquake. However, after less than a month, he had to go back to

media oversight of local government disaster relief work. When local officers were the only ones who knew what was going on, they were tempted to abuse their power and exploit victims.[101]

This problem is not specific to China. Any system involving distributing large sums of money is vulnerable to dishonest contractors and corrupt officials. As every government learns after disasters, money corrupts.[102]

2. The Samaritan's Dilemma

The Samaritan's Dilemma[103] haunts governmental aid. According to James Buchanan's definition, the Samaritan's Dilemma is created when the government makes direct payments to individuals after a disaster, giving them incentives not to take protective measures or purchase insurance but instead to rely on the government to bail them out.[104] Even if the government promises ex ante not to provide such relief, the promise is not credible, because it will be in everyone's interests to offer such relief after a disaster has struck.[105] Therefore, more government bailouts may cause more disaster losses because people are less likely to take precaution measures.[106]

The Whole-Nation System poses the Samaritan's Dilemma. Often, a government bailout is motivated by an admirable humanitarian impulse that spurs redistributing wealth to those who have suffered loss from those who have been spared.[107] Political concerns are also important under the Whole-Nation System. Although the Chinese central government has no re-election constraints, governmental performance in providing relief during and after the disaster impacts the support for the authorities at all

---

Beijing to deal with other national affairs and paid less and less attention to earthquake relief after the emergency work was accomplished. See *Wen Jiabao: Please Remember the Great Sichuan Earthquake*, China News (May 24, 2008), available at http://www.chinanews.com/gn/news/2008/05-24/1260997.shtml.

[101]   Justin Yifu Lin, Demystifying the Chinese Economy 198 (2012).

[102]   *Katrina: Four Years Later—Fraud, Corruption Cases Continue* (Federal Bureau of Investigation, 2009), available at http://www.fbi.gov/news/stories/2009/september/katrina_090109.

[103]   It is based on an old parable of the traveler from Samaria who helped a stranger whom he found beaten and robbed by the side of the road.

[104]   James M. Buchanan, *The Samaritan's Dilemma* in Altruism, Morality and Economic Theory 71 (E.S. Phelps ed., 1972).

[105]   James M. Buchanan, *The Samaritan's Dilemma* in Altruism, Morality and Economic Theory 71 (E.S. Phelps ed., 1972).

[106]   Tom Baker, *On the Genealogy of Moral Hazard*, 75 Texas Law Review 237 (1996).

[107]   George L. Priest, *Government Insurance versus Market Insurance*, 28 The Geneva Papers on Risk and Insurance—Issues and Practice 71 (2003).

levels.[108] However, this humanitarian and political action ignores the fact that in some cases, the effect of the redistribution will encourage future loss-causing activities that would not otherwise have been undertaken.[109] Some pure forms of government bailout, including ad hoc direct payment and compensation funds, provide insufficient incentives for risk prevention and loss mitigation.[110] Therefore, individuals will be less inclined to protect against disaster when they believe the government will bail them out, which increases the magnitude of loss for the whole nation.

In addition, the Chinese people have historically had a strong inclination to rely on a governmental bailout in the wake of a catastrophe. Under the Whole-Nation System, the government is committed to restoring social and economic order after a disaster. Thanks to such governmental commitment, individuals' personal experiences, and media reports of past catastrophes, it is rational for the Chinese people to fail to take adequate protective measures. Many residents admit that they are exposed to catastrophe risks, but they seldom transfer risks through insurance because they believe the government will bail them out when catastrophes happen.[111] According to an empirical study on property and causality insurance in five Chinese provinces, there is a negative correlation between the amount of government relief and residents' investment in prevention measures such as purchasing insurance.[112]

In short, the Samaritan's Dilemma, which reduces individuals' incentives to invest in protective measures and leads to further government bailouts, is a challenge for the Whole-Nation System.[113]

---

[108]   Egor Lazarev, Anton Sobolev, Irina V. Soboleva and Boris Sokolov, *Trial by Fire: A Natural Disaster's Impact on Support for the Authorities in Rural Russia*, 66 World Politics 641 (2014).

[109]   George L. Priest, *Government Insurance versus Market Insurance*, 28 The Geneva Papers on Risk and Insurance—Issues and Practice 71 (2003).

[110]   Roger Van den Bergh and Michael Faure, *Compulsory Insurance of Loss to Property Caused by Natural Disasters: Competition or Solidarity?*, 29 World Competition 25 (2006).

[111]   He Wang, Research on Catastrophe Risk Insurance Mechanisms 5 (2013).

[112]   L. Tian and Y. Zhang, *Influence Factors of Catastrophe Insurance Demand in China—Panel Analysis in a Case of Insurance Premium Income of Five Provinces [Woguo Juzai Baoxian Xuqiu Yingxiang Yinsu Shizheng Yanjiu: Jiyu Wusheng Bufen Baofei Shouru Mianban Yanjiu]*, 26 Wuhan University of Technology (Social Science Edition) [Wuhan Ligong Daxue Xuebao (Shehui Kexue Ban)] 175 (2013).

[113]   James M. Buchanan, *The Samaritan's Dilemma* in Altruism, Morality and Economic Theory 71 (E.S. Phelps ed., 1972).

Table 1.5   Comparison between some supporting provinces and supported counties in per capita fiscal expenditure before the 2008 earthquake (RMB)

Supporting provinces or municipalities	Per capita fiscal expenditure (2007)	Supported counties	Per capita fiscal expenditure (2007)
Hunan	2,135	Lixian County	4,209
Jilin	3,237	Heishui County	4,149
Anhui	2,033	Songpan County	4,107
Jiangxi	2,072	Xiaojin County	3,056

Source: Ni, Feng, Zhang, Yue, and Yu, Tongzhou, *Wenchuan Dadizhen Duikou Zhiyuan Chubu Yanjiu [Preliminary Research on Counterpart Aid of Wenchuan Earthquake]*, 7 Jingji Guanli Yu Yanjiu [Research on Economics and Management] 55–62 (2009).

3. The regressive effects of counterpart aid

Counterpart aid is a mechanism within the Whole-Nation System under which the central government organizes a special "pooling" of catastrophe risk across the country. It relies on the maxim that "one province helps one significantly affected county", in theory requiring the richer provinces— not all provinces—to help disaster-affected areas and contribute more in disaster relief. However, when the disaster is a widespread catastrophe and many counties are affected, this "one province helps one significantly affected county" arrangement frequently leads to the situation in which not-that-rich provinces have to help reconstruct richer (by per capita fiscal expenditure) counties. For example, Hunan Province was responsible for the reconstruction of Lixian County after the Great Sichuan Earthquake. However, the per capita fiscal expenditure of Hunan Province was only RMB 2,135 in 2007; of Sichuan Province, RMB 2,165; and of Lixian County, RMB 4,209.[114] Similar per capita disparities existed in Jilin, Anhui, and Jiangxi provinces (Table 1.5).

Counterpart aid aims to realize the goal of "common prosperity" through richer provinces helping poorer ones. In practice, however, sometimes poorer provinces have to help richer ones. Adding to the problem is that there are no legal provisions regulating counterpart aid, which means

[114] Feng Ni, Yue Zhang and Tongzhou Yu, *Wenchuan Dadizhen Duikou Zhiyuan Chubu Yanjiu [Preliminary Research on Counterpart Aid of Wenchuan Earthquake]*, 7 Jingji Guanli Yu Yanjiu [Research on Economics and Management] 55 (2009).

that the obligations of supporting provinces are unclear and arbitrary and decided at the discretion of the central government.[115]

Sometimes, the counterpart aid arrangement has not followed the principle of equality or efficiency but has resulted from political pressure. For example, after the Xinjiang Riot in 2009, one year later, 19 provinces were required to supply counterpart aid to Xinjiang Uygur Autonomous Region (hereinafter "Xinjiang") for its reconstruction and development just as was required for post-disaster restoration and reconstruction of the Great Sichuan Earthquake.[116] The per capita GDP of some supporting provinces was lower than that of Xinjiang.[117] Three years later, more supporting provinces, including Hunan, Henan, and Shanxi, were still lower than Xinjiang in per capita GDP. From this perspective, the Whole-Nation System perpetuates inequality, and the regressive effects of counterpart aid are more severe.

Even when poor areas suffer from catastrophe losses, unfairness may arise. For example, after the 2008 Great Sichuan Earthquake, Shandong Province, which ranked in the top three in GDP in 2007, was responsible for the reconstruction of BeiChuan County and donated RMB 12 billion (about $2 billon).[118] The per capita fiscal expenditure of Shandong Province was RMB 2,415 in 2007, while in Beichuan County it was RMB 2,299.[119] Five years later, however, after receiving the counterpart aid, BeiChuan County

---

[115]   L. Zhao and Y.J. Jiang, *Analysis of Local Government Coordinated Assistance Modes [Difang Zhengfu Duikou Zhiyuan Moshi Fenxi]*, 2 Journal of ChengDu University (Social Science Edition) [ChengDu Daxue Xuebao (Sheke Ban)] 4, 25 (2009).

[116]   Xlong Wenzhao and Tian Yan, *Research on the Legalization of Partner Assistance Policy in Xinjiang [Duikou Yuanjiang Zhengce De Fazhihua Yanjiu]*, 31 Journal of Xinjiang Normal University (Social Science) [Xinjiang Shifan Daxue Xuebao (Shehui Kexue Ban)] 12 (2010).

[117]   Per capital GDP of Jiang Xi province (RMB 21,288) and An Hui province (RMB 20,002) are lower than Xinjiang (RMB 23,159). See Xlong Wenzhao and Tian Yan, *Research on the Legalization of Partner Assistance Policy in Xinjiang [Duikou Yuanjiang Zhengce De Fazhihua Yanjiu]*, 31 Journal of Xinjiang Normal University (Social Science) [Xinjiang Shifan Daxue Xuebao (Shehui Kexue Ban)] 12 (2010).

[118]   L. Zhao and Y.J. Jiang, *Analysis of Local Government Coordinated Assistance Modes [Difang Zhengfu Duikou Zhiyuan Moshi Fenxi]*, 2 Journal of ChengDu University (Social Science Edition) [ChengDu Daxue Xuebao (Sheke Ban)] 4, 25 (2009).

[119]   Feng Ni, Yue Zhang, and Tongzhou Yu, *Wenchuan Dadizhen Duikou Zhiyuan Chubu Yanjiu [Preliminary Research on Counterpart Aid of Wenchuan Earthquake]*, 7 Jingji Guanli Yu Yanjiu [Research on Economics and Management] 55 (2009).

was much more prosperous than many Shandong Province counties.[120] In 2012, the per capita fiscal expenditure in Shandong Province was around RMB 5,900, while in Beichuan County it was around RMB 8,000.[121]

In fact, counterpart aid does not eliminate risk but only distributes the burden of disaster losses across the taxpayers of supporting provinces. Therefore, counterpart aid emerges as an arguably unfair arrangement for the residents of the supporting provinces because the supporting provinces may treat their own residents "less favorably" than those in disaster areas. Under China's political selection system, in which the central government has the final word rather than the residents,[122] governors of supporting provinces have strong incentives to deal with disaster-affected areas more favorably than they do with their own residents because the counterpart aid is politically favored by the central government.

4. Lack of risk financing

Risk financing is regarded as one of the three pillars of risk management and classically requires those who face risks to pay for coverage through risk-based premiums.[123] Ex ante insurance with risk-based premiums provides incentives to accumulate reserves and mitigate losses before disasters, while ex post government aid may reduce incentives to reserve funds or carry out mitigation activities.[124] The Whole-Nation System pays little attention to, and indeed lacks, the capability of risk financing through ex ante insurance markets to compensate victims, instead focusing on ex post relief. To illustrate, of the compensation for the Great Sichuan Earthquake, only $300 million of the losses were covered by insurance companies, i.e., only 0.3 percent of the total losses incurred.[125]

---

[120]   He Wang, Research on Catastrophe Risk Insurance Mechanisms 86 (2013).
[121]   Annual Fiscal Report of Beicuan County (2012), available at http://www.beichuan.gov.cn/html/2013/gsgg_0529/10227.html?cid=11; Annual Fiscal Report of Shandong Province (2012), available at http://www.mof.gov.cn/zhuantihuigu/2013ssyshb/201302/t20130219_733742.html.
[122]   Lian Zhou, *Administrative Subcontract [Xizheng Fabaozhi]*, 6 Shehui [Society] 1 (2014).
[123]   F. Outreville, Theory and Practice of Insurance 45–64 (1998); R. Thoyts, Insurance Theory and Practice 286–295 (2010). (The other two are risk assessment (also known as risk analysis) and risk control).
[124]   Dwight Jaffee and Thomas Russell, *The Welfare Economics of Catastrophe Losses and Insurance*, 38 The Geneva Papers on Risk and Insurance—Issues and Practice 469 (2013).
[125]   *Establishing Catastrophe Insurance System Faces Acceleration*, China Youth Daily (March 14, 2011), available at http://zqb.cyol.com/html/2011-03/14/nw.D110000zgqnb_20110314_1-05.htm?div=-1.

Furthermore, aspects of the Whole-Nation System, such as counterpart aid, "crowd out" the establishment and development of risk financing markets by depressing the demand of individuals for catastrophe insurance. Individuals are less likely to purchase insurance to prefinance their potential disaster losses. According to general international experiences, a catastrophe insurance system is often established within two years after a disaster occurs. For example, in 1944, after the 1942 earthquake in New Zealand, an earthquake insurance system was established.[126] California's experience was similar.[127] However, five years after the 2008 Great Sichuan Earthquake, a catastrophe insurance market had not yet been officially established in China. The Whole-Nation System, which played a powerful role in earthquake relief, at least partially accounts for this.

5.   Other challenges and problems
Besides the above problems of the Whole-Nation System, there are other potential challenges.

Disaster relief imposes burdens on public budgets and may hinder economic growth. In smaller and developing countries, a catastrophe event can result in higher public deficits and debt.[128] Although China is now the second largest economy in the world, the cost of the Whole-Nation System still imposes a considerable burden on the public budget. For example, the government spent approximately $166 billion after the 2008 earthquake in restoration and reconstruction.[129] Despite that, there is no consensus on whether disaster relief depresses or in fact enhances economic growth.[130] A two-period equilibrium model indicates that direct payment of disaster relief funds may depress rather than enhance economic growth because

---

[126]   D.J. Dowrick, *Damage and Intensities in the Magnitude 7.8 1929 Murchison, New Zealand, Earthquake*, 27 Bulletin of the New Zealand National Society for Earthquake Engineering 190 (1994); D.J. Dowrick, *Earthquake Risk Reduction Actions for New Zealand*, 36 New Zealand Society for Earthquake Engineering 249 (2003).

[127]   Dwight Jaffee and T. Russell, *Behavioral Models of Insurance: The Case of the California Earthquake Authority*, University of California-Berkeley Working Paper (2000).

[128]   Swiss Re, *Disaster Risk Financing: Reducing the Burden on Public Budgets*, Swiss Re Focus Report (2008).

[129]   The total fiscal revenues were around $1 trillion in 2008. See Jiang Lingling et al., *People's Republic of China: Providing Emergency Response to Sichuan Earthquake*, Technical Assistance Consultant's Report to Ministry of Civil Affairs, P.R. China and Asian Development Bank (2008).

[130]   Xian Xu and Jiawei Mo, *The Impact of Disaster Relief on Economic Growth: Evidence from China*, 38 The Geneva Papers on Risk and Insurance—Issues and Practice 495 (2013).

disaster relief related to the loss of capital and the substitution effect of direct transfer payment depresses post-disaster labor supply.[131] Such effects of disaster relief on growth have been tested using panel data on 31 Chinese provinces, municipalities, and autonomous regions over the period 2004–2010.[132] The extensive use of the military in disaster relief has been a double-edged sword because it could potentially displace the development of an effective civilian-based disaster management system in the future.[133]

In addition, the Whole-Nation System leads to an unstable budget because it operates as an ad hoc compensation tool for victims, which may dampen economic growth.

The misappropriation of funds or goods also contributes to the negative impact of the Whole-Nation System. According to a survey report in earthquake-affected areas after the 2008 Great Sichuan Earthquake, the Ministry of the Civil Affairs was in charge of medical aid money, while the Ministry of Health was responsible for patients' treatment. This mismatch of rights and obligations between different departments increases transaction costs but also leads to the misuse of capital and medical resources.[134]

E.   A Short Conclusion

After considering the structure of the Whole-Nation System, how it works, and its problems, we may see that the Whole-Nation System has indeed played a valuable role in dealing with catastrophe losses in China. More important insights for the Whole-Nation System, however, suggest opportunities for improvement. Summarized in Figure 1.4, the System's pre-disaster measures as ex ante mitigation actions are efficient policies to address catastrophe losses, as are its post-disaster measures in emergency response.

Figure 1.4 also shows the problems of the Whole-Nation System. What the System did in pre-disaster measures is too little, not too much. For

---

[131]   F. Barry, *Government Consumption and Private Investment in Closed and Open Economies*, 21 Journal of Macroeconomics 93 (1999); T. McDermott, F. Barry and R. Tol, *Disasters and Development: Natural Disasters, Credit Constraints and Economic Growth*, ESRI Working Paper No. 411 (2011).

[132]   Xian Xu and Jiawei Mo, *The Impact of Disaster Relief on Economic Growth: Evidence from China*, 38 The Geneva Papers on Risk and Insurance—Issues and Practice 495 (2013).

[133]   Jian Zhang, *The Military and Disaster Relief in China: Trends, Drivers and Implications* in Disaster Relief in the Asia Pacific: Agency and Resilience (Minako Sakai et al. eds., 2014).

[134]   He Wang, Research on Catastrophe Risk Insurance Mechanisms 92 (2013).

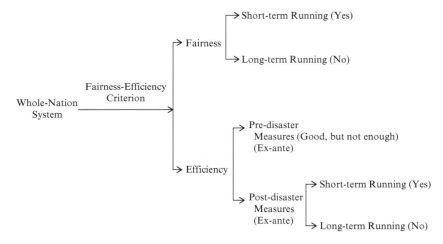

Figure 1.4   Evaluation of the Whole-Nation System

actions post-disaster, there should be a close examination of the methods of victims' compensation and counterpart aid.

The Whole-Nation System, in this limited context, seems to have performed reasonably well, but it is far from a totally efficient, sustainable, and long-term catastrophe risk management system. Although private insurance must deal with various market failures in covering catastrophe risks, it is difficult to say that the Whole-Nation System is superior to catastrophe insurance. To deal with the risk of natural disasters more effectively in China, there is an urgent need for the reform of the Whole-Nation System to integrate it with insurance that could finance underlying catastrophe risks and offer other advantages. Two methods will be proposed. One is the government stimulation of private catastrophe insurance (the following section and Chapter 2, 3 and 4), and the other is government sponsored reinsurance to finance and support primary catastrophe insurance due to market failure (Chapter 5). Such a reform could improve social welfare and advance social justice.

IV.   GOVERNMENT STIMULATION OF PRIVATE CATASTROPHE INSURANCE MARKETS FOR DISASTERS

Having reviewed the performance of the Whole-Nation System—its achievements and problems—we are now ready to explore its reform and transition. Compared with the government-run Whole-Nation System, insurance is traditionally recognized as an important tool to deal with catastrophic disasters, "through risk pooling and risk shifting, but also risk reduction and risk management".[135] Moreover, many law and economics scholars favor insurance as a private-market mechanism for distributing catastrophe risk, especially when compared to government-provided compensation.[136] For example, Jaffe and Russell,[137] Kunreuther,[138] Epstein,[139] Priest,[140] and Kaplow[141] argue that insurance is better equipped to deal with catastrophe risks than the government due to its advantages of lower transaction costs, lower adverse selection, and greater efficiency. Michel-Kerjan et al. posit that as the level of economic development improves within a state, public and private catastrophe insurance should gradually develop (Figure 1.5).[142] According to this simplified view, China should be moving from step 2 to step 3, consistent with the observation that catastrophe insurance could be a necessary and proper complement to the Whole-Nation System. Before the discussion of insurance's role in addressing catastrophe losses,[143] however, the starting point should be how to create a catastrophe insurance market in China where, for whatever reason, it has still failed to emerge.

---

[135] Omri Ben-Shahara and Kyle D. Logue, *Outsourcing Regulation: How Insurance Reduces Moral Hazard*, 111 Michigan Law Review 197 (2012).

[136] Michael Faure and Klaus Heine, *Insurance Against Financial Crises?*, 8 NYU Journal of Law & Business 117 (2011).

[137] Dwight Jaffe and Thomas Russell, *Catastrophe Insurance, Capital Markets, and Uninsured Risks*, 62 Journal of Risk and Insurance 205 (1997).

[138] Howard Kunreuther, *The Case for Comprehensive Disaster Insurance*, 11 Journal of Law and Economics 133 (1968).

[139] Richard A. Epstein, *Catastrophe Responses to Catastrophe Risks*, 12 Journal of Risk and Uncertainty 287 (1996).

[140] George L. Priest, *The Government, the Market, and the Problem of Catastrophe Loss*, 12 Journal of Risk and Uncertainty 219 (1996).

[141] Louis Kaplow, *Incentives and Government Relief for Risk*, 4 Journal of Risk and Uncertainty 167 (1991).

[142] "Level 1: Very limited funding from central government; heavy reliance on donors; Level 2: Mainly ex-post funding from the central government; Level 3: Some coverage by insurance systems; government is still the main funding source; Level 4: Significant (re)insurance penetration; government supplements by allocating catastrophic risk capital". See Erwann Michel-Kerjan et al., *Catastrophe Financing for Governments: Learning from the 2009–2012 MultiCat Program in Mexico*, OECD Working Papers on Finance, Insurance and Private Pensions, No. 9 (2011).

[143] This will be discussed in Chapter 2.

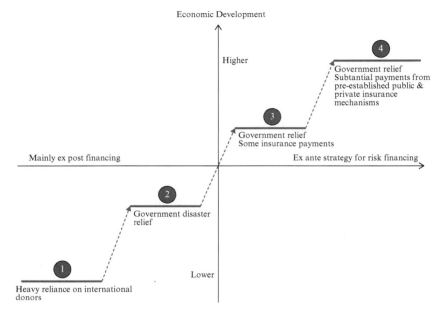

Source: Erwann Michel-Kerjan, et al., Catastrophe Financing for Governments: Learning from the 2009–2012 MultiCat Program in Mexico, OECD Working Papers on Finance, Insurance and Private Pensions, No. 9, (2011).

Figure 1.5   A simplified view of governments' responses to financial management of natural disasters

A.   The Immature Catastrophe Insurance Market in China

In 1959, the Chinese government closed all domestic insurance companies, and the insurance business as a whole virtually disappeared from 1959 to 1978.[144] After the Third Plenum of the Eleventh Party Congress, the process of institutional transformation towards a market-oriented economy began, and commercial insurance agencies also resumed doing business. In 1979, the State Council approved the "Notice on Restoration of the Domestic Insurance Business and Strengthening of the Insurance Agency" (*Guanyu Huifu Guonei Baoxian Yewu He Jiaqiang Baoxian Jigou De Tongzhi*), which allowed insurers to underwrite property insurance, vehicle insurance, marine insurance, and life insurance. Since 1989, the insurance industry has

---

[144]   He Wang, Research on Catastrophe Risk Insurance Mechanisms 87–89 (2013).

been one of the fastest growing industries in China, with nominal premium income growing at an annual average rate of 30 percent.[145] China's insurance market ranked as the fourth largest in the world in 2013.[146]

Catastrophe insurance, however, has walked a much bumpier road than other lines in the insurance industry. In 1987, the Ministry of Civil Affairs launched the agricultural insurance pilot projects in Heilongjiang, Fujian, and Jiangsu provinces.[147] It is worth noting that these agricultural insurance policies covered catastrophic risks such as droughts and floods. However, these pilot projects were closed after 12 years' experimentation.[148] In 1996, floods, typhoons, and other natural disasters were all prescribed as exclusions by regulators in property insurance policies.[149] Thus, these catastrophic exposures were excluded from coverage under standard-form policies. As a result, in the 1998 Yangtze River flood, insurance covered only about 1.25 percent of economic losses—$500 million coverage out of a total loss of $40 billion.[150] In 2000, the China Insurance Regulatory Commission (CIRC)—the newly established insurance regulatory agency—prohibited insurers from underwriting earthquake insurance policies without its permission.[151] As a result, only 0.3 percent of the total losses were covered by insurance companies in the 2008 Great Sichuan Earthquake.[152]

This earthquake underscored the need for catastrophe insurance. In

---

[145]   Bingzheng Chen, Sharon Tennyson, Maoqi Wang and Haizhen Zhou, *The Development and Regulation of China's Insurance Market: History and Perspectives*, 17 Risk Management and Insurance Review 241 (2013).

[146]   Xinhua News Agency (2013), available at http://www.gov.cn/xinwen/2014-07/09/content_2714415.htm.

[147]   Luo Guo Liang, *Process and Experiences: Disaster Mitigation of New China for Sixty Years*, 5 Beijing Social Science 73 (2009).

[148]   Peijun Shi, Di Tang, Jing Liu, Bo Chen and M.Q. Zhou, *Natural Disaster Insurance: Issues and Strategy of China* in Asian Catastrophe Insurance 79 (2008).

[149]   People's Banks of China, *Notice on the Term, Rate and Provision Interpretation of* <Property Basic Insurance> *and* <Property Comprehensive Insurance> *[Guanyu Yinfa* <Caichan Baoxian Jibenxian>he <Caichan Baoxian Zonghexian>Tiaokuan, Feilv ji Tiaokuan Jieshi de Tongzhi]* (1996), available at http://law.baidu.com/000570240002206ed685b5a29b1feb0fd7ae84890045.html.

[150]   Ministry of the Civil Affairs of the PRC, *The Statistics Report of the Civil Affairs Development [Minzheng Shiye Fazhan Baogao]* (1998), available at http://cws.mca.gov.cn/article/tjbg/200801/20080100009419.shtml.

[151]   CIRC, *The Notice of Regulating Approving Procedure of Extended Earthquake Insurance in Enterprise Property Insurance Policy [Guanyu Qiye Chaichan Baoxian Yewu Bude Kuozhan Chengbao Dizhen Fengxian de Tongzhi]* (2000), available at: http://www.chinalawedu.com/news/1200/21829/21838/2197 9/2006/3/pa1359122815161360024334-0.htm. The basis for this Regulation was concern for the solvency of insurance companies in the face of earthquake risks.

[152]   *Establishing Catastrophe Insurance System faces acceleration*, China

2012, one of the main topics in the Fourth National Finance Working Conference was establishing a system of catastrophe insurance.[153] In 2013, the "Regulation on Agriculture Insurance" was promulgated, which included government-subsidized agricultural catastrophe insurance.[154] For example, Article 8 provides that the central government should establish subsidized agricultural catastrophe insurance, and local governments are encouraged to follow and contribute to the pool; Article 17 prescribes that any insurance company that wants to underwrite an agricultural insurance policy must include catastrophe insurance coverage. The Regulation includes a provision on catastrophe insurance; however, there are no details on how the government will establish the system, how much it will subsidize policyholders, how insurers should prepare for the catastrophe insurance coverage, and many other technical questions. In 2014, catastrophe insurance program trials were begun in Shenzhen, in the Pearl River Delta (a densely populated metropolitan area and also one of the world's most disaster-prone regions), and in the Chuxiong region in the southwestern province of Yunnan, which is prone to earthquakes.[155] These pilot programs have yet to be evaluated. Overall, catastrophe insurance in China is still at a very immature stage and cannot yet meet the rising demand for disaster relief. It needs proper government intervention.

B.   Theoretical Framework for Developing a Catastrophe Insurance Market: Overcoming the Market-Government Dichotomy and Looking for Customized Solutions

Generally speaking, there are three major theoretical frameworks for analysing government responsibilities in market intervention: laissez-faire theory, also called the market-friendly view; public-interest theory, also called the developmental-state view; and market-enhancing

---

Youth Daily (March 14, 2011), available at http://zqb.cyol.com/html/2011-03/14/nw.D110000zgqnb_20110314_1-05.htm?div=-1.

[153]   The National Finance Working Conference is the supreme financial conference in China, which decides on the most important financial issues, such as establishing the China Insurance Regulation Commission (CIRC) (2012), available at http://finance.ce.cn/sub/2011/jrgzhy.

[154]   Decree No. 619 of the State Council of the People's Republic of China (2013), available at: http://www.gov.cn/zwgk/2012-11/16/content_2268392.htm.

[155]   *China says testing catastrophe insurance system*, Reuters (August 20, 2014), available at http://www.businessinsurance.com/article/20140820/NEWS04/14082 9990?AllowView=VDl3UXk1T3hDUFNCbkJiYkY1TDJaRUt0ajBRV0ErOVV HUT09#.

theory.[156] How to develop an insurance industry generally falls into one of these three camps.

1. Laissez-faire theory

The laissez-faire theory of government policy believes that the efficient allocation of resources within the economy can be achieved through the market mechanism; even when markets alone are insufficient, other private-sector organizations will suffice, and the outcome still remains more efficient than government intervention.[157] According to this theory, private insurance markets can achieve efficient market equilibrium, and government's role is limited to providing a fair legal environment for market transactions.[158] In such an analysis, calling for government intervention in providing catastrophe insurance would be viewed as opportunistic "rent-seeking" attempts of special-interest groups to secure an ex ante wealth transfer from taxpayers.[159]

In many respects, this theoretical framework has merit. In the absence of distortion-inducing government intervention, the outcome of the private insurance market may remain more efficient. For example, regarding rate regulation of catastrophe insurance, government often depresses insurance rates in response to political pressure. As a result, the premium cannot reflect the real risk. Insurers can no longer cover the

---

[156] See Masahiko Aoki, Kevin Murdock and Masahiro Okuno-Fujiwara, *Beyond the East Asian Miracle: Introducing the Market-Enhancing View* in The Role of Government in East Asian Economic Development: Comparative Institutional Analysis (Masahiko Aoki, Hyung-Ki Kim and Masahiro Okuno-Fujiwara eds., 1998); C.M. Lewis and K.C. Murdock, *Alternative Means of Redistributing Catastrophic Risk in a National Risk Management System* in The Financing of Catastrophe Risk 74 (K. Froot ed., 1999); J.D. Cummins and O. Mahul, Catastrophe Risk Financing in Developing Countries: Principles for Public Intervention 84–85 (2009); Veronique Bruggeman, Compensating Catastrophe Victims: A Comparative Law and Economics Approach 193–195 (2010).

[157] J.D. Cummins and O. Mahul, Catastrophe Risk Financing in Developing Countries: Principles for Public Intervention 84–85 (2009); Veronique Bruggeman, Compensating Catastrophe Victims: A Comparative Law and Economics Approach 193–195 (2010).

[158] Masahiko Aoki, Kevin Murdock and Masahiro Okuno-Fujiwara, *Beyond the East Asian Miracle: Introducing the Market-Enhancing View* in the Role of Government in East Asian Economic Development: Comparative Institutional Analysis (Masahiko Aoki, Hyung-Ki Kim and Masahiro Okuno-Fujiwara eds., 1998).

[159] J.D. Cummins and O. Mahul, Catastrophe Risk Financing in Developing Countries: Principles for Public Intervention 84–85 (2009).

variable cost of providing insurance or recoup their initial investments in providing service, and this creates an availability crisis.[160] The UK's private flood insurance scheme, developed about half a century ago, offers an example of how largely private markets could work.[161] Private insurance companies compensate victims in the case of flood damage. Although the British government sets standards and rules for flood protection, land-use, and flood warning, it promises to guarantee the independence of privately run compensation "schemes" according to a gentleman's agreement that defines the rights and duties of state and industry.[162] The UK private flood insurance scheme has remained largely unchanged since its emergence in 1961 and has proven to be efficient and sustainable.[163]

However, the laissez-faire theory, in which government is considered as a black box usually unsuccessfully solving problems of markets, is largely inappropriate for dealing with the practical problems of economic development and social reform in transitional economies.[164] The theory confronts the "paradox of the adjusting state": on the one hand, "the government is required to withdraw from interventions into economic activities and to perform a more passive role"; on the other hand, "economic transition and development usually require nimble and robust political authorities to implement and enforce the new market-oriented policy directives due to existing market imperfections".[165] Making the government more effective so that it can solve new challenges and perform new roles in facilitat-

---

[160]   C.M. Lewis and K.C. Murdock, *Alternative Means of Redistributing Catastrophic Risk in a National Risk Management System* in The Financing of Catastrophe Risk 75 (K. Froot ed., 1999).

[161]   Michael Huber, *Insurability and Regulatory Reform: Is the English Flood Insurance Regime Able to Adapt to Climate Change?*, 29 The Geneva Papers on Risk and Insurance—Issues and Practice 169 (2004).

[162]   Michael Huber, *Insurability and Regulatory Reform: Is the English Flood Insurance Regime Able to Adapt to Climate Change?*, 29 The Geneva Papers on Risk and Insurance—Issues and Practice 169 (2004).

[163]   Michael Huber, *Insurability and Regulatory Reform: Is the English Flood Insurance Regime Able to Adapt to Climate Change?*, 29 The Geneva Papers on Risk and Insurance—Issues and Practice 169 (2004).

[164]   J. Ahrens, *Governance in the Process of Economic Transformation* in System Transformation in Comparative Perspective: Affinity and Diversity in Institutional, Structural and Cultural Patterns (M. Jovanovic, L. Dalgon, O. Yeon-Chean, S. Park and B. Seliger eds., 2007).

[165]   J. Ahrens, *Governance in the Process of Economic Transformation* in System Transformation in Comparative Perspective: Affinity and Diversity in Institutional, Structural and Cultural Patterns (M. Jovanovic, L. Dalgon, O. Yeon-Chean, S. Park and B. Seliger eds., 2007).

ing private-sector coordination is of utmost importance for economic transformation and social reform.[166] However, it is not explicitly included in the laissez-faire theoretical framework.

Catastrophe insurance started only recently in China. In the aftermath of the 2008 Great Sichuan Earthquake, however, a catastrophe insurance market has still not yet been officially established.[167] Only recently has the government decided to intervene: some pilot programs of catastrophe insurance in the cities of Shenzhen and Ningbo have been implemented since 2014.[168]

Clear imperfections in the insurance market in China make it is easy to perceive why catastrophe insurance has not been quickly established since the Great Sichuan Earthquake. From the supply side, insurers face a lot of obstacles to underwriting catastrophe insurance policies, such as lack of catastrophe data to identify, quantify, and estimate the chance of disasters and to set premiums; relatively low capacity of the insurance industry; limited access to international reinsurance and capital markets; regulatory obstacles; and so forth. From the demand side, consumers do not always behave rationally and maximize their expected utility to protect themselves from catastrophe losses. Their demand for catastrophe

---

[166] J. Ahrens, *Governance in the Process of Economic Transformation* in System Transformation in Comparative Perspective: Affinity and Diversity in Institutional, Structural and Cultural Patterns (M. Jovanovic, L. Dalgon, O. Yeon-Chean, S. Park and B. Seliger eds., 2007).

[167] It is still not formally available, although there is coverage for some natural disasters in some property insurance policies underwritten by the People's Insurance Company of China (PICC). According to general international experiences, a catastrophe insurance system is often established within two years after a catastrophe disaster. For example, after the 1999 Marmara Earthquake in Turkey, with the support of the World Bank and Turkish government, the Turkish Catastrophe Insurance Pool (TCIP) was created in 2000. See Johann-Adrian von Lucius, *A Reinsurance Perspective on the Turkish Catastrophe Insurance Pool* in Catastrophe Risk and Reinsurance: A Country Risk Management Perspective 217 (Eugene N. Gurenko ed., 2004); D.M. Jaffee and T. Russell, *Behavioral Models of Insurance: the Case of the California Earthquake Authority*, University of California-Berkeley Working Paper (2000).

[168] According to the latest news, in July 2014, the Government of Shenzhen bought a catastrophe insurance policy from PICC on behalf of the residents of Shenzhen. This catastrophe insurance framework includes three different parts: the first part is the government catastrophe assistance insurance, which was bought by the Shenzhen municipal government to supply the basic assistance for all residents; the second part is a catastrophe fund; and the third part is private catastrophe insurance. See http://xw.sinoins.com/2014-07/10/content_120490.htm.

Similar catastrophe insurance arrangement is in Ningbo City. See http://insurance.hexun.com/2014-11-04/170012090.html.

insurance has also been blunted by expectations of a government bailout. For these reasons, if the laissez-faire theory were followed, the catastrophe insurance market would fail to develop soon in China and also fail to play its potential role in mitigating and financing catastrophe losses.

2. Public-interest theory

The public-interest theory of government policy contests laissez-faire theory. It holds that government intervention should correct market failures (such as externalities, imperfect competition, moral hazard, and adverse selection). By solving the suboptimal allocation of resources, government intervention can improve social welfare.[169] According to this theoretical framework, government intervention is often seen as a substitute for or complement to imperfect coordination in the private markets.[170] Advocates consider government to have better information and judgment than private insurers and to be able to guide markets wisely.[171] For developing countries where market failure is more pervasive, government intervention is more urgently needed. Limitations on insurance market infrastructure, lack of liquid capital markets, information asymmetries, and other imperfections associated with the catastrophe insurance market necessitate a government role in developing a private market, according to supporters of public-interest theory.[172]

Even in the United States, where Americans have long believed in the gospel of free markets, government intervention in the management of private-sector risks is nothing new.[173] The government and private lenders want savers to feel confident in their banks and credit unions so that the system of saving and lending is maintained.[174] As a result, the

---

[169]   J.D. Cummins and O. Mahul, Catastrophe Risk Financing in Developing Countries: Principles for Public Intervention 84–85 (2009).

[170]   C.M. Lewis and K.C. Murdock, *Alternative Means of Redistributing Catastrophic Risk in a National Risk Management System* in The Financing of Catastrophe Risk 74 (K. Froot ed., 1999).

[171]   Masahiko Aoki, Kevin Murdock and Masahiro Okuno-Fujiwara, *Beyond the East Asian Miracle: Introducing the Market-Enhancing View* in The Role of Government in East Asian Economic Development: Comparative Institutional Analysis (Masahiko Aoki, Hyung-Ki Kim and Masahiro Okuno-Fujiwara eds., 1998).

[172]   J.D. Cummins and O. Mahul, Catastrophe Risk Financing in Developing Countries: Principles for Public Intervention 84–85 (2009).

[173]   David A. Moss, When All Else Fails: Government as the Ultimate Risk Manager 17 (2004).

[174]   Sophie M. Korczyk, *Insuring the Uninsurable: Private Insurance Markets and Government Intervention in Cases of Extreme Risks*, NAMIC Issue Analysis (2005).

deposit insurance system and the Federal Deposit Insurance Corporation were established in order to prevent the type of mass banking panic that crippled the American financial system in the early 1930s and to protect depositors and maintain confidence in the banking system.[175] The US government and consumers wanted to ensure an adequate level of farming activity even in the face of potentially catastrophic weather risks, so that the nation was not excessively dependent on imported food.[176] As a result, the Federal Crop Insurance Program, which is administered by the Federal Crop Insurance Corporation within the US Department of Agriculture, was established.[177] The government and employers wanted people to continue to live and work in certain geographic areas, e.g., the Borough of Manhattan in New York City and Silicon Valley in California, where there are risks, though infrequent, of terrorist attacks, floods, hurricanes, and earthquakes.[178] As a result, the Terrorism Risk Insurance Act was passed in November 2002, which provided government reinsurance of terrorist risks, and the National Flood Insurance Program was established and sponsored by the government. The California Earthquake Authority is also quasi-public.[179] When private insurers have failed to emerge or withdrawn underwriting, in consideration of the public interest, the US government has chosen to secure continuing insurance markets.

Although market failure and the need for economies of scale are common justifications for the government to intervene in an insurance market, these are not, alone, sufficient to improve the allocation of resources.[180] On the contrary, many public insurance programs have been shown to be inefficient. The US National Flood Insurance Program is an example. The federal government is unlikely to implement general insurance underwriting principles, such as risk segregation, price discrimination to control adverse

---

[175]   Tom Baker and David Moss, *Government as Risk Manager* in New Perspectives on Regulation 87 (David Moss and John Cisternino eds., 2009).

[176]   Veronique Bruggeman, Compensating Catastrophe Victims: A Comparative Law and Economics Approach 193–195(2010).

[177]   Barry J. Barnett, *US Government Natural Disaster Assistance: Historical Analysis and a Proposal for the Future*, 23 Disasters 139 (1999).

[178]   Sophie M. Korczyk, *Insuring the Uninsurable: Private Insurance Markets and Government Intervention in Cases of Extreme Risks*, NAMIC Issue Analysis (2005).

[179]   Dwight Jaffee and Thomas Russell, *The Welfare Economics of Catastrophe Losses and Insurance*, 38 The Geneva Papers on Risk and Insurance—Issues and Practice 469 (2013).

[180]   Sophie M. Korczyk, *Insuring the Uninsurable: Private Insurance Markets and Government Intervention in Cases of Extreme Risks*, NAMIC Issue Analysis (2005).

selection, and mechanisms to control moral hazard, because of political pressure and lack of market experience.[181] As a result, the non-risk-based premiums tend to distort the market price and encourage policyholders to overinvest in risky areas and to take inadequate steps to mitigate losses.

In contrast with laissez-faire theory, public-interest theory requires government to act as a substitute for coordination in the private markets. However, public-interest theory is not suitable for China's market-oriented transition and reform. At present, China is still struggling through the transition from a centrally planned economy to a market economy. No independent business sectors or free markets existed until 1978.[182] During the transition process, China adopted a gradual dual-track approach, rather than shock therapy, and moved slowly to a well-functioning market economy.[183] The experience of the dual-track approach shows that the more energetic power is coming from private companies. They, as opposed to state-owned enterprises, played a significant role in economic growth. The existence of state-owned enterprises is mainly due to political consideration, to avoid the collapse of old-priority industries, and to maintain rents for those who may be negatively affected by the shift towards a market system.[184] Nonetheless, the dual-track approach as the transitional arrangement can hardly serve as the permanent foundation for further development, and China needs to move from the dual tracks of planned economy and market economy to a single-track market economy.[185]

If China adopted public-interest theory, the threat of a reunion of the government and the state-owned insurance enterprises would become reality and we would see the dual-track approach backslide to the single-track planned economy.[186] The case of a collaborative rural house insurance

---

[181]   Adam F. Scales, *A Nation of Policyholders: Governmental and Market Failure in Flood Insurance*, 26 Mississippi College Law Review 3 (2006).

[182]   Feng Wang and Haitao Yin, *A New Form of Governance or the Reunion of the Government and Business Sector? A Case Analysis of the Collaborative Natural Disaster Insurance System in the Zhejiang Province of China*, 15 International Public Management Journal 429 (2013).

[183]   Justin Yifu Lin, *Demystifying the Chinese Economy*, 46 The Australian Economic Review 259 (2013).

[184]   Justin Yifu Lin, *Demystifying the Chinese Economy*, 46 The Australian Economic Review 259 (2013).

[185]   Justin Yifu Lin, Demystifying the Chinese Economy 178 (2012).

[186]   Feng Wang and Haitao Yin, *A New Form of Governance or the Reunion of the Government and Business Sector? A Case Analysis of the Collaborative Natural Disaster Insurance System in the Zhejiang Province of China*, 15 International Public Management Journal 429 (2013).

system in the Zhejiang Province is a vivid example of such challenges. In this program, initiated in 2006, insurance products are primarily designed by the government and marketed through administrative mobilization instead of market channels run by the People's Insurance Company of China (PICC; Zhejiang branch). Even though the provisions of "The Notice of Developing Rural House Insurance" (2006) require that the operations of the market should follow the principles: (i) insurance companies operate based on market mechanisms and (ii) government promotes insurance coverage through subsidization.[187] Legislation states clearly that the insurance company, rather than the government, should be the primary operator as an independent market entity. In fact, however, the government intervenes in an all-encompassing manner, and there are no incentives for insurance companies to market products; additionally, since only People's Insurance Company of China Property and Casualty Company Limited ("PICC P&C") is allowed to participate in such business, this company free rides on government efforts and enjoys monopolistic benefits.[188] As a result, a well-functioning and effective rural house insurance market is far from being established.[189]

3. Market-enhancing theory

The market-enhancing theory of government policy is different from both laissez-faire theory and public-interest theory and takes a middle position. It recognizes that market failure leads to suboptimal allocation of resources and thus suggests that the government can facilitate more efficient coordination in the private sector and enhance the development of the private market. It also suggests that government should not substitute or replace the private sectors and should especially avoid creating permanent new government institutions to substitute for private insurers.[190] This theory was first applied by Lewis and Murdock to managing catastrophe disaster

---

[187] Zhejiang Provincial Government, *The Notice of Developing Rural House Insurance* (November, 2006), available at http://www.zjmz.gov.cn/il.htm?a=si&id=4028e4812c33ef74012c9c000b4e0582.

[188] Feng Wang and Haitao Yin, *A New Form of Governance or the Reunion of the Government and Business Sector? A Case Analysis of the Collaborative Natural Disaster Insurance System in the Zhejiang Province of China*, 15 International Public Management Journal 429 (2013).

[189] Feng Wang and Haitao Yin, *A New Form of Governance or the Reunion of the Government and Business Sector? A Case Analysis of the Collaborative Natural Disaster Insurance System in the Zhejiang Province of China*, 15 International Public Management Journal 429 (2013).

[190] J.D. Cummins and O. Mahul, Catastrophe Risk Financing in Developing Countries: Principles for Public Intervention 84–85 (2009).

risk as early as the 1990s.[191] It has been widely discussed by scholars and international financial institutions, such as the World Bank.[192]

Market-enhancing theory represents a new kind of government intervention that helps facilitate the creation of private markets and assists private insurers to solve market failures. In contrast to laissez-faire theory, this framework looks for the role of government in achieving more efficient market equilibrium. This is particularly true for transitional and developing countries where private institutions are more limited and presently unable to solve all market failures.[193] In contrast to public-interest theory—the traditional government intervention—it promotes the decentralized decision making of private insurers and avoids creating bureaucracy. In other words, the provision of catastrophe insurance policies should be left to private insurers.[194] This is particularly important for transitional countries where, under the impact of a planned economy, government is still powerful and might be tempted to meddle in micro-management of insurance activities. As is well known, the establishment of efficient catastrophe insurance programs depends on the integration of four key components: risk assessment, risk pooling, risk segregation, and

---

[191]   See C.M. Lewis and K.C. Murdock, *The Role of Government Contracts in Discretionary Reinsurance Markets for Natural Disasters*, 63 Journal of Risk and Insurance 567 (1996); C.M. Lewis and K.C. Murdock, *Alternative Means of Redistributing Catastrophic Risk in a National Risk Management System* in The Financing of Catastrophe Risk 51 (K. Froot, ed., 1999).

[192]   See, e.g. Eugene Gurenko, Rodney Lester, Olivier Mahul and Serap Oguz Gonulal, Earthquake Insurance in Turkey: History of the Turkish Catastrophe Insurance Pool (2006); Thomas Hellmann, Kevin Murdock and Joseph Stiglitz, *Financial Restraint and the Market Enhancing View* in IEA Conference Volume Series 2559 (1998); J.D. Cummins and O. Mahul, Catastrophe Risk Financing in Developing Countries: Principles for Public Intervention 84–85 (2009); Veronique Bruggeman, Compensating Catastrophe Victims: A Comparative Law and Economics Approach 193–195 (2010); World Bank, *Country Insurance: Reducing Systemic Vulnerabilities in Latin America and the Caribbean*, Report 43066-LAC (2007); World Bank, *India: National Agricultural Insurance Scheme: Market-Based Solutions for Better Risk Sharing*, Report 39353 (2007); World Bank, *The World Bank Group's Catastrophe Risk Financing Products and Services*, World Bank Catastrophe Risk Insurance Working Group (2008).

[193]   Masahiko Aoki, Kevin Murdock and Masahiro Okuno-Fujiwara, *Beyond the East Asian Miracle: Introducing the Market-Enhancing View* in The Role of Government in East Asian Economic Development: Comparative Institutional Analysis (Masahiko Aoki, Hyung-Ki Kim and Masahiro Okuno-Fujiwara eds., 1998).

[194]   C.M. Lewis and K.C. Murdock, *Alternative Means of Redistributing Catastrophic Risk in a National Risk Management System* in The Financing of Catastrophe Risk 74 (K. Froot ed.,1999).

control of moral hazard.[195] The government should create a legal environment to enhance the private development of these components rather than get involved in the details of setting premiums and underwriting policies.

This theoretical framework has many merits. It helps establish an effective and sustainable catastrophe risk financing system with collaboration between government and private insurers. The government avoids acting as a direct insurer, but it can act as a reinsurer or lender of last resort when private underwritings are unavailable or inadequate.[196] More importantly, this theoretical framework encourages public-private partnership, which could help facilitate access to international reinsurance and capital markets to generate affordable products for domestic insurers.[197] In practice, this market-enhancing theoretical framework has already attracted the attention of international financial institutions like the World Bank, which guides government intervention in catastrophe risk markets of low- and middle-income countries.[198] In 2000, the Turkish Catastrophe Insurance Pool (TCIP) was established with the assistance of the World Bank in the aftermath of the Marmara Earthquake.[199] The TCIP is managed by a board of seven members; insurance companies cede 100 percent of all risk to the pool; the adjustment of claims is done by independent loss adjustors; and the full risk capital requirements of the TCIP are funded through commercial reinsurance, including Milli Re and Munich Re.[200] The TCIP sold more than 4.8 million policies set at market-based premium rates in 2012, compared to 600,000 covered households when the pool was set up, and penetration rate rose to 29 percent nationwide.[201] A similar catastrophe pool is being developed in Romania, and there are

---

[195]   J.D. Cummins and O. Mahul, Catastrophe Risk Financing in Developing Countries: Principles for Public Intervention 87 (2009).
[196]   J.D. Cummins and O. Mahul, Catastrophe Risk Financing in Developing Countries: Principles for Public Intervention 87 (2009).
[197]   J.D. Cummins and O. Mahul, Catastrophe Risk Financing in Developing Countries: Principles for Public Intervention 87 (2009).
[198]   J.D Cummins and O. Mahul, Catastrophe Risk Financing in Developing Countries: Principles for Public Intervention 8–9 (2009).
[199]   Johann-Adrian von Lucius, *A Reinsurance Perspective on the Turkish Catastrophe Insurance Pool* in Catastrophe Risk and Reinsurance: A Country Risk Management Perspective 217 (Euguene N. Gurenko ed., 2004).
[200]   Johann-Adrian von Lucius, *A Reinsurance Perspective on the Turkish Catastrophe Insurance Pool* in Catastrophe Risk and Reinsurance: A Country Risk Management Perspective 217 (Euguene N. Gurenko ed., 2004).
[201]   Turkish Catastrophe Insurance Pool (TCIP) English Annual Reports (2012), available at http://www.tcip.gov.tr/content/annualReport/2012_Annual_Report_DASK.pdf.

now even multiple-country regional organizations, such as the Caribbean Catastrophe Risk Insurance Facility.[202]

In addition, market-enhancing theory consists of government regulation of private insurance. Think of all of the regulatory techniques that apply to private insurers in the United States, European Union and other countries, including entry controls, minimum capital requirements, other types of solvency regulation, regulation of forms, and prohibitions against unfair trade practices. If a private catastrophe insurance market takes off in China, some or all of those types of regulation in the interest of solvency and consumer protection are warmly welcomed (and the CIRC already does that for other types of insurance in China).

Like many other areas of the economy, the developing catastrophe insurance market in China can draw lessons from the accumulated achievements of the ongoing socioeconomic reforms—the so-called Chinese miracle.[203] The essence of China's "miracle" is consistent with market-enhancing theory, rather than laissez-faire theory or public-interest theory.[204] The Chinese approach stresses the fundamental role of the market in resource allocation but also expects the government to play a facilitating role by addressing externalities, coordination, and many other market-failure issues.[205]

---

[202]   J.D. Cummins and O. Mahul, Catastrophe Risk Financing in Developing Countries: Principles for Public Intervention 93 (2009).

[203]   Elizabeth J. Perry, *Growing Pains: Challenges for a Rising China*, 143 Daedalus 5 (2014).

[204]   Simply speaking, the main experiences of China's "miracle" development are: (i) well-functioning competitive markets, which are the precondition for developing a country's industries because only with such a market can prices reflect scarcities of production in the economy and propel firms to enter industries; (ii) a proactive, facilitating government, which is equally important because, for transitional countries, the government should seize and capture the advantages of late-comers through playing a role in information collection, coordination, and compensation for externalities. See Justin Yifu Lin, *Demystifying the Chinese Economy*, 46 The Australian Economic Review 259 (2013).

[205]   The Chinese approach strives to institute a new hybrid system, which, indeed, is the market-enhancing approach. China does not follow shock therapy— a policy recommendation based on orthodox economic theory—promoting economic liberalization and privatization, which is more or less the application of laissez-faire theory. The failure of Russian's transition, a typical example of shock therapy, however, reveals that the liberalization-cum-privatization approach does not automatically bring about efficient and sustainable market structures. China's success and Russia's failure prove that in developed countries the government should perhaps interfere less, but for transitional countries, a minimalist government is not optimal. On the other hand, a strong, proactive and facilitating state could start and enhance genuine market-oriented reforms, because the leaders of

Considering the low-probability but high-consequence nature of catastrophe risk and the currently immature condition of the insurance market and economic development, the market-enhancing theory could be the proper guiding theory to develop a catastrophe insurance market in China. In addition, market-enhancing theory conforms to the practice of catastrophe insurance pilot programs in China. The market-enhancing theory emphasizes the importance of local information and suggests that decentralized private agents can use locally available information to come up with market-based solutions that are significantly more efficient than those that could be imposed by a central authority.[206]

Nonetheless, potential challenges should be closely examined when applying the market-enhancing theory in China. China has not transitioned into a market economy despite decades of reform. Thus, government may easily slide into intervening in microbusiness activities in the name of public interest, due to path dependency.[207] The case of the collaborative rural house insurance system, which mainly covers the losses caused by natural disasters in the Zhejiang Province of China, is a reflection of how

---

China recognize that only genuine market reforms and sustained economic growth that benefits all strata of society—not just the political elite and big business—can supply their political legitimacy. This direction differs from the public-interest theory, which requires government to act as a substitute for private sectors. See Justin Yifu Lin, *New Structural Economics: A Framework for Rethinking Development*, 26 The World Bank Research Observer 193 (2011). Justin Yifu Lin, the former Chief Economist and Senior Vice President of the World Bank (2008–2012), based on China's experiences of development, raised new structure theory as a theoretical explanation. See Justin Yifu Lin, Demystifying the Chinese Economy (2012); Justin Yifu Lin, New Structural Economics: A Framework for Rethinking Development (2012); Justin Yifu Lin, The Quest for Prosperity: How Developing Economies Can Take Off (2012). See also J. Ahrens, *Governance in the Process of Economic Transformation* in System Transformation in Comparative Perspective: Affinity and Diversity in Institutional, Structural and Cultural Patterns (M. Jovanovic, L. Dalgon, O. Yeon-Chean S. Park and B. Seliger eds., 2007).

[206] C.M. Lewis and K.C. Murdock, *Alternative Means of Redistributing Catastrophic Risk in a National Risk Management System* in The Financing of Catastrophe Risk 77 (K. Froot ed., 1999).

[207] V. Nee and Y. Cao, *Path Dependent Societal Transformation: Stratification in Hybrid Mixed Economies*, 28 Theory and Society 799 (1999): "The theory of path dependence states that history matters for contemporary institutional development." See S.J. Liebowitz and S.E. Margolis, *Path Dependence*, 1 Encyclopedia of Law and Economics 981 (2000): "Based on state-centered accounts of transitions, the long-standing vertical ties linking government bureaus with economic actors perpetuate patterns of government and business relationships deeply entrenched in economy and society."

the government and the insurance companies work with each other in a way similar to how they would operate in a planned economy.[208]

4. A short conclusion

By examining laissez-faire theory, public-interest theory, and market-enhancing theory, I suggest that market-enhancing theory might be a proper theoretical guide for the development of a catastrophe insurance market in China. As Ronald Coase said, "satisfactory views on policy can only come from a patient study of how, in practice, the market, firms, and governments handle the problem of harmful effects".[209] Such study indicates that devising a private catastrophe insurance market system and enhancing the role of insurance in addressing catastrophe losses has become an urgent challenge in China.

C.   Principles for Government Intervention in the Catastrophe Insurance Market

Based on the above analysis of market-enhancing theory and the experiences of China's transitional development, I propose a catastrophe insurance market-enhancing framework to coordinate the role of government and market in catastrophe risk management in China. The content of this framework has three pillars:

- Sustaining a strong and capable government
- Enhancing the market, while neither supplanting nor retarding it
- Legalizing the relationship between government and market to prevent government from undermining well-functioning market operations

To efficiently apply the catastrophe insurance market-enhancing framework, and to help establish a well-functioning catastrophe insurance market and disaster compensation system,[210] I further propose three

---

[208]   Feng Wang and Haitao Yin, *A New Form of Governance or the Reunion of the Government and Business Sector? A Case Analysis of the Collaborative Natural Disaster Insurance System in the Zhejiang Province of China*, 15 International Public Management Journal 429 (2013).

[209]   Ronald Coase, *The Problem of Social Cost*, 3 Journal of Law and Economics 1 (1960).

[210]   Based on the comparative analysis of types of systems in different countries, the ideal standard consists of: (i) financially sustainable, (ii) adequate policies for preventing and mitigating risks, and (iii) the provision of affordable insurance

principles to facilitate government intervention in the catastrophe insurance market:

- Principle 1. Government should help solve market failures in catastrophe insurance and secure insurers' business operations using market mechanisms, rather than creating new government institutions to supplant private solutions.
- Principle 2. Government may establish a mandatory catastrophe insurance system through legislation and help solve the affordability problem at the same time.
- Principle 3. Government should reform the Whole-Nation System to avoid crowding out private insurance and enhance the collaboration between the insurance industry and the government.

1.   Government responsibilities under Principle 1

Due to insurability limits with catastrophe risk and the high potential losses of catastrophe exposures, insurers lack the capacity and appetite to sufficiently cover all such losses.

(a)   The predictability condition of insurability Insurability of catastrophe risk is an important consideration for insurers when they decide whether to underwrite policies. If the risk is not insurable, insurers will have no appetite, even if they have financial capacity to write more business. Generally speaking, there are two agreed-upon requirements for insurability: (i) the insurer must have the ability to identify, quantify, and estimate the chances of disasters and the resulting losses; (ii) the insurer must have the ability to set and collect appropriate premiums for catastrophe risks.[211] To help fulfill the insurability requirements, governments can assist insurers with risk assessment, mapping risk zones, information flows, and so on.

Risk assessment is used to discover the underlying actuarial costs

---

with low management expenses to a broad public in hazard-prone areas. See J.R. Skees and B.J. Barnett, *Conceptual and Practical Considerations for Sharing Catastrophic/Systemic Risks*, 21 Review of Agricultural Economics 424 (1999); Youbaraj Paudel, *A Comparative Study of Public–Private Catastrophe Insurance Systems: Lessons from Current Practices*, 37 The Geneva Papers on Risk and Insurance—Issues and Practice 257 (2012).

[211]   Howard Kunreuther and E.O. Michel-Kerjan, *Climate Change, Insurability of Large-Scale Disasters, and the Emerging Liability Challenge*, 155 University of Pennsylvania Law Review 1795 (2007); Howard Kunreuther and R.J. Roth, Sr, Paying the Price: the Status and Role of Insurance against Natural Disasters in the United States 27–38 (1998).

of catastrophe risk and help set accurate risk-adjusted premiums.[212] It requires data collection, catastrophe risk modeling, and other technical support. Government, for example, has the advantage in data collection of natural disasters due to its disaster-relief experiences. Mapping risk zones can be regarded as a means of risk assessment because it depicts and summarizes specific hazard risks of properties or zones.[213] It could best be conducted through collaboration between governments and insurers.[214] For example, European countries were required to prepare flood hazard and flood risk maps before 2014 to comply with the 2007 European Flood Risk Management Directive.[215] Insurance businesses are heavily dependent on information flows, which include information flowing from policyholder to intermediaries, from intermediaries to insurers, and from insurers to reinsurers.[216] If information cannot flow smoothly and accurately, it will not only increase transaction costs but make it difficult for insurers/reinsurers to identify, quantify, and correctly estimate the chances of disasters occurring and the resulting losses.

(b)   Increasing the capacity of the insurance industry Due to the highly correlated and aggregate nature of natural disasters, the potential losses from catastrophe risks are large and uncertain. At this stage, China's insurers may still lack sufficient financial capacity to fully cover catastrophe losses. The total net capital of China's property insurance companies is much lower than the total amount of losses caused by natural disasters.

As early as 1992, following Hurricane Andrew, US insurers realized that outside capital was needed to supplement the industry's capacity.[217] The government can enhance the use of capital markets as a financing

---

[212]   J.D. Cummins and O. Mahul, Catastrophe Risk Financing in Developing Countries: Principles for Public Intervention 76–77 (2009).

[213]   Youbaraj Paudel, *A Comparative Study of Public–Private Catastrophe Insurance Systems: Lessons from Current Practices*, 37 The Geneva Papers on Risk and Insurance—Issues and Practice 257–285 (2012).

[214]   Youbaraj Paudel, *A Comparative Study of Public–Private Catastrophe Insurance Systems: Lessons from Current Practices*, 37 The Geneva Papers on Risk and Insurance—Issues and Practice 257 (2012).

[215]   J. Van Alphen, F. Martini, R. Loat, R. Slomp and R. Passchier, *Flood Risk Mapping in Europe, Experiences and Best Practices*, 2 Journal of Flood Risk Management 285 (2009).

[216]   J.D. Cummins and O. Mahul, Catastrophe Risk Financing in Developing Countries: Principles for Public Intervention 31–32 (2009).

[217]   R. King, *Hurricanes and Disaster Risk Financing Through Insurance: Challenges and Policy Options*, Congressional Research Service, R132825 (2005).

Table 1.6   Capital of main Chinese property insurers compared to natural disaster losses (US\$ billions)

	2007	2008	2009	2010
Net capital of main insurers	5.5	5.1	6.9	9.0
Natural disaster losses	38.1	189.5	40.1	86.1

Source: Yearbook of China Insurance (2008–2011).

mechanism for insurance through equity stakes by outside investors in insurance companies, bond issuances, securitization, etc. This is particularly important for the Chinese government because its insurance/reinsurance market and capital markets are still in their infancy. To make the enhancement of capital markets work, the government should facilitate its development by eliminating market barriers, reducing transaction costs, and ensuring the rule of law for those markets.

In addition, government may act as a reinsurer to help increase the capacity of the insurance industry. Effective and sustainable catastrophe risk financing solutions require collaboration between government and private insurers, especially in the case of extreme catastrophes when the private reinsurance market "hardens".[218] The designers of the government's role as reinsurer may learn from the well-functioning Turkish Catastrophe Insurance Pool (TCIP). When earthquake losses in Turkey exceed \$80 million, the excess losses are transferred to the reinsurance market.[219] The Turkish government covers "losses that would exceed the overall claims paying capacity of the TCIP, which is currently sufficient to withstand a 1-in-350 year earthquake".[220] The TCIP is supported by the

---

[218]   For example, in France and Japan, where catastrophe coverage is mandatory, all catastrophe insurance policies written by private insurers are reinsured by the government-run reinsurance company, which essentially serves as guarantor for all policies written. In the US, under the Terrorism Risk Insurance Act of November 2002, the federal government agreed to provide a kind of reinsurance "backstop" for terrorism losses. See Milton Nektarios, *A Catastrophe Insurance System for the European Union*, 5 Asia-Pacific Journal of Risk and Insurance, Article 6 (2011); Thomas Russell and Jeffrey E. Thomas, *Government Support for Terrorism Insurance*, 15 Connecticut Insurance Law Journal 183 (2008).

[219]   Johann-Adrian von Lucius, *A Reinsurance Perspective on the Turkish Catastrophe Insurance Pool* in Catastrophe Risk and Reinsurance: A Country Risk Management Perspective 217 (Eugene N. Gurenko ed., 2004).

[220]   Turkish Catastrophe Insurance Pool, available at http://www.gfdrr.org/sites/gfdrr.org/files/documents/DFI_TCIP__Jan11.pdf.

World Bank, which provides financial and technical assistance.[221] As the first national catastrophe insurance pool in World Bank-client countries, the TCIP is a good model for China.

Besides acting as a reinsurer of last resort to increase the capacity of the insurance industry, the government could sell industry-level, excess-of-loss contracts for insured disaster losses as a last resort, as proposed by Lewis and Murdock.[222]

(c)   Promoting risk classification and encouraging risk-based premiums
Risk classification and risk-based premiums are the heart of a healthy insurance business.[223] Only by segregating policyholders into different risk pools can insurers charge different premiums for different pools, reduce adverse selection, and incentivize risk reduction by policyholders, and thus make profits.[224] However, government and government-run catastrophe insurance systems are unlikely to implement this basic principle when faced with political pressures[225] because risk classification is not always compatible with social solidarity objectives that promote equal treatment of all citizens.[226] Therefore, the government should neither create new institutions to supplant private solutions nor depress premiums on insurance policies. Instead, it should allow insurers to set the premiums to reflect risks. Even if there are concerns about the affordability of

---

[221]   Turkish Catastrophe Insurance Pool, available at http://www.gfdrr.org/sites/gfdrr.org/files/documents/DFI_TCIP__Jan11.pdf.

[222]   Excess-of-loss contracts are "designed to complement existing private-sector insurance and reinsurance mechanisms by covering only layers of reinsurance currently unavailable in the private market ... These contracts, which are equivalent to call-spread options written on an industry index of disaster losses, would be auctioned to qualified insurance companies and would carry a maturity of one year". See C.M. Lewis and K.C. Murdock, *Alternative Means of Redistributing Catastrophic Risk in a National Risk Management System* in The Financing of Catastrophe Risk 80 (K. Froot ed., 1999).

[223]   Kenneth S. Abraham, *Efficiency and Fairness in Insurance Risk Classification*, 71 Virginia Law Review 403 (1985).

[224]   Ronen Avraham, Kyle D. Logue and Daniel Schwarcz, *Understanding Insurance Antidiscrimination Laws*, 87 Southern California Law Review 195 (2014).

[225]   Youbaraj Paudel, *A Comparative Study of Public–Private Catastrophe Insurance Systems: Lessons from Current Practices*, 37 The Geneva Papers on Risk and Insurance—Issues and Practice 257 (2012) ("In general, the fully public private insurance systems have not integrated risk-based premiums and financial incentives for mitigation").

[226]   J.D. Cummins and O. Mahul, Catastrophe Risk Financing in Developing Countries: Principles for Public Intervention 76–78 (2009).

catastrophe insurance, it is better to take measures from the perspective of consumer demand (such as insurance vouchers, discussed below) than to depress insurers' incentives to underwrite policies and distort risk signals provided by actuarially based premiums. As a Chinese proverb has it, "you can't expect the horse to run fast when you don't let it graze". Government cannot expect insurers to underwrite policyholders' risks if it does not let them make profits.

(d) Encouraging disaster mitigation activities Disaster mitigation activities can benefit both policyholders and insurers because they decrease the costs of the catastrophe insurance system in the long run. Catastrophe insurance systems in different countries encourage mitigation activities.[227] The government should encourage disaster mitigation policies that include but are not limited to conducting natural disaster risk investigation and zoning; establishing natural disaster monitoring systems and early-warning systems; implementing building code standards; pushing forward natural disaster prevention projects; and investing in public protection infrastructure.[228]

2. Government responsibilities under Principle 2
Due to the low-probability but high-consequence nature of catastrophe risk, supply of and demand for catastrophe insurance is low. Principle 2 attempts to solve this issue by establishing a mandatory system.

(a) Establishing a mandatory catastrophe insurance system Probably one of the most debated issues in establishing catastrophe insurance systems is whether those systems should be mandatory. Unlike private insurers, the government or lawmakers have power to compel consumers to participate in insurance programs, no matter whether those programs are government-run or private-run insurance.[229] Opponents and proponents both advance

---

[227] Youbaraj Paudel, *A Comparative Study of Public–Private Catastrophe Insurance Systems: Lessons from Current Practices*, 37 The Geneva Papers on Risk and Insurance—Issues and Practice 257 (2012).

[228] Office of National Commission for Disaster Reduction, P.R. China, *China's Natural Disaster Risk Management* in Improving the Assessment of Disaster Risks to Strengthen Financial Resilience 121 (A Special Joint G20 Publication by the Government of Mexico and the World Bank, 2012); Youbaraj Paudel, *A Comparative Study of Public–Private Catastrophe Insurance Systems: Lessons from Current Practices*, 37 The Geneva Papers on Risk and Insurance—Issues and Practice 257 (2012).

[229] David A. Moss, When All Else Fails: Government as the Ultimate Risk Manager 49–50 (2004).

arguments to justify their positions.[230] Generally speaking, the potential challenges associated with the mandatory catastrophe insurance system include (i) violating freedom of contract, (ii) cross-subsidization, and (iii) anticompetition. In contrast, the potential benefits generally include (i) correcting irrational behaviors to justify infringements on freedom of contract, (ii) managing adverse selection to justify cross-subsidization, (iii) enhancing national solidarity to justify anticompetition, and (iv) promoting damage mitigation. Although there are several (potential) challenges to a mandatory insurance system, the advantages often justify this system.

More importantly, due to the low-probability nature of catastrophe risks and the reliance on government bailouts under the Whole-Nation System, Chinese consumers have little incentive to purchase catastrophe insurance products voluntarily. Like consumers everywhere, Chinese citizens do not always behave rationally and maximize their expected utility to protect themselves from catastrophe losses by buying insurance. Prior to a disaster, consumers believe that natural disasters will not happen to them.[231]

---

[230]   An overview of the debate is as follows: Kunreuther proposed a mandatory model as early as 1968. This opinion was repeated by other scholars, especially after Hurricane Katrina, and also received support from European scholars. See Howard Kunreuther, *The Case for Comprehensive Disaster Insurance*, 11 Journal of Law and Economics 133 (1968); Howard Kunreuther, *Has the Time Come for Comprehensive Natural Disaster Insurance?* in On Risk and Disaster: Lessons from Hurricane Katrina 175, 175 (Ronald Daniels, Donald F. Kettl and Howard Kunreuther eds., 2006); Howard Kunreuther and Mark Pauly, *Rules rather than Discretion: Lessons from Hurricane Katrina*, 33 Journal of Risk & Uncertainty 101, 102–104 (2006); Anastasia Telesetsky, *Insurance as a Mitigation Mechanism: Managing International Greenhouse Gas Emissions through Nationwide Mandatory Climate Change Catastrophe Insurance*, 27 Pace Environmental Law Review 691 (2010). European scholarship includes: Reimund Schwarze and Gert G. Wagner, *In the Aftermath of Dresden: New Directions in German Flood Insurance*, 29 The Geneva Papers on Risk and Insurance—Issues and Practice 154, 163 (2004); Michael G. Faure, *Insurability of Damage Caused by Climate Change: A Commentary*, 155 University of Pennsylvania Law Review 1875 (2007); Michael Faure and Veronique Bruggeman, *Catastrophic Risks and First-Party Insurance*, 15 Connecticut Insurance Law Journal 1 (2008). On the other hand, some literature indicates the potential dangers to mandatory insurance system, such as regulatory paternalism, anticompetition, and overgeneralization. See Anthony I. Ogus, *Regulatory Paternalism: When Is It Justified?* in Corporate Governance in Context: Corporations, States, and Markets in Europe, Japan and the US 303, 303 (Klaus J. Hopt et al. eds., 2005); Scott E. Harrington, *Rethinking Disaster Policy*, 23 Regulation 40 (2000).

[231]   Shlomo Benartzi and Richard Thaleer, *Myopic Loss Aversion and the Equity Premium Puzzle*, 110 Quarterly Journal of Economics 73 (1995).

(b)   Solving the affordability problem When deciding to establish a mandatory catastrophe insurance system, there is an obvious tension between a high-risk-based premium and affordability, given the severity and spatial correlation of catastrophe losses.[232] The pricing debate over the National Flood Insurance Program in the United States is a vivid example of this tension. On March 21, 2014, President Obama signed the Homeowner Flood Insurance Affordability Act of 2014, which responded to political complaints of "unaffordability" by repealing and softening certain provisions of the Biggert-Waters Flood Insurance Reform Act of 2012, which moved toward premiums reflecting risk.[233] The affordability problem is even more severe in China as a developing country.

A government may provide basic coverage for residents. The basic coverage of catastrophe losses as part of social safety net programs could be justified because the public social insurance is insufficient in China.[234] However, this coverage should be implemented via ex ante catastrophe insurance rather than compensating victims directly via the ex post Whole-Nation System, which faces many problems. From this perspective, outsourcing coverage via purchasing insurance may be more efficient.

Ideally, coverage of catastrophe losses can be regarded as an integrated system consisting of different layers. Government coverage is just the first layer, and the remaining layers supplied by private insurers can cover broader exposures. In this instance, to solve the affordability problem of poor people, the government could supply insurance vouchers rather than traditional direct-premium subsidies to insurance buyers. One example of a direct-premium subsidy is an arrangement where 50 percent of the risk-based premium is paid by policyholders and the rest is paid by the government.[235] Direct-premium subsidies that depress premiums "tend to have highly distortional implications for the insurance markets and risk management behavior of the policyholders".[236] The US Crop Insurance Program and India's National Agricultural Insurance Scheme are two

---

[232]   Carolyn Kousky and Howard Kunreuther, *Addressing Affordability in the National Flood Insurance Program*, 01 Journal of Extreme Events 1450001 (2014).
[233]   FEMA, Homeowner Flood Insurance Affordability Act: Overview (2014), available at http://www.fema.gov/media-library-data/1396551935597-4048b68f6d695a6eb6e6e7118d3ce464/HFIAA_Overview_FINAL_03282014.pdf.
[234]   J.D. Cummins and O. Mahul, Catastrophe Risk Financing in Developing Countries: Principles for Public Intervention 81–83 (2009).
[235]   J.D. Cummins and O. Mahul, Catastrophe Risk Financing in Developing Countries: Principles for Public Intervention 82–83 (2009).
[236]   J.D. Cummins and O. Mahul, Catastrophe Risk Financing in Developing Countries: Principles for Public Intervention 82–83 (2009).

examples demonstrating the imperfection of direct-premium subsidy.[237] The problems of direct-premium subsidies include criticisms that (i) they are untargeted and available to all policyholders, without distinction between low-income households and high-income households; (ii) they tend to become permanent, even though the government initially introduces them as temporary subsidies; (iii) they put an increasing fiscal burden on the government as the scope of the subsidy creeps up; and (iv) they mainly benefit policyholders located in high-risk zones.[238]

In contrast, insurance vouchers that recipients can use like cash but only to buy qualifying insurance could help secure risk-based premiums, ensure that insurers play their correct role, and solve the affordability problem for low-income households.[239] Unlike a direct-premium subsidy, means-tested insurance vouchers ensure that recipients use the funds for obtaining insurance, without distorting the insurance premium reflecting risk.[240] The amount of the voucher can be determined by using a sliding scale based on annual family income, and a voucher program could be coupled with mitigation requirements so as to reduce future disaster losses.[241] In the United States, there are several kinds of well-functioning voucher programs, such as the Housing Choice Voucher Program, the Food Stamp Program, the Low Income Home Energy Assistance Program, and the Universal Service Fund.[242]

---

[237]   J.D. Cummins and O. Mahul, Catastrophe Risk Financing in Developing Countries: Principles for Public Intervention 82–83 (2009).

[238]   J.D. Cummins and O. Mahul, Catastrophe Risk Financing in Developing Countries: Principles for Public Intervention 82–83 (2009).

[239]   Carolyn Kousky and Howard Kunreuther, *Addressing Affordability in the National Flood Insurance Program*, 01 Journal of Extreme Events 1450001 (2014); Howard Kunreuther and Erwann, Michel-Kerjan, *Enhancing Post-Disaster Economic Recovery: How Improved Flood Insurance Mechanisms Can Help* in Current Research Project Synopses 71 (2014); Erwann Michel-Kerjan, Jeffrey Czajkowski and Howard Kunreuther, *Could Flood Insurance be Privatized in the United States? A Primer*, 39 The Geneva Papers on Risk and Insurance—Issues and Practice 651 (2015); Howard Kunreuther, *Reducing Losses from Catastrophic Risks through Long-Term Insurance and Mitigation*, 75 Social Research: An International Quarterly 905 (2008).

[240]   Howard Kunreuther, *Reducing Losses from Catastrophic Risks through Long-Term Insurance and Mitigation*, 75 Social Research: An International Quarterly 905 (2008).

[241]   Carolyn Kousky and Howard Kunreuther, *Addressing Affordability in the National Flood Insurance Program*, 01 Journal of Extreme Events 1450001 (2014).

[242]   Carolyn Kousky and Howard Kunreuther, *Addressing Affordability in the National Flood Insurance Program*, 01 Journal of Extreme Events 1450001 (2014).

3. Government responsibilities under Principle 3

Principle 3 reflects the need to reform the current catastrophe disaster compensation arrangements under the Whole-Nation System to enhance the collaboration between the insurance industry and the government. Therefore, some arrangements of the Whole-Nation System should be reformed—such as counterpart aid—to avoid crowding out private insurance.

How to coordinate the Whole-Nation System and market-based catastrophe insurance depends on whether relief under the Whole-Nation System crowds out private insurance transactions. Basic rational choice theory tells us that if individuals treat ad hoc relief from the Whole-Nation System as a substitute for insurance, they will underinsure or fail to insure at all.[243] According to the empirical studies on different post-disaster relief schemes comparing Austria, where governmental relief is certain but incomplete, and Germany, where governmental relief is uncertain but more complete, the results show that "expected governmental relief has a strong crowding-out effect on insurance demand and this effect is even more pronounced when governmental relief is more certain".[244] In other words, the government relief scheme of Austria has a stronger crowding-out effect for market-based insurance because of its certainty. Unfortunately, China's Whole-Nation System is more like the Austrian than the German model. It means that, under the Whole-Nation System, even introducing market-based catastrophe insurance may not play an expected and desired role. Counterpart aid in the form of direct disaster grants to affected households is an example. This form of assistance is likely to crowd out private insurance markets, according to some researchers.

For example, an empirical study suggests that the direct disaster grants of the US Federal Emergency Management Agency (FEMA) have a

---

[243] Paul A. Raschky, Reimund Schwarze, Manijeh Schwindt and Ferdinand Zahn, *Uncertainty of Governmental Relief and the Crowding out of Flood Insurance*, 54 Environmental and Resource Economics 179 (2013); Tracy Lewis and David Nickerson, *Self-Insurance against Natural Disasters*, 16 Journal of Environmental Economics and Management 209 (1989); Louis Kaplow, *Incentives and Government Relief for Risk*, 4 Journal of Risk and Uncertainty 167 (1991); Mary Kelly and Anne E. Kleffner, *Optimal Loss Mitigation and Contract Design*, 70 Journal of Risk and Insurance 53 (2003); Carolyn Kousky, Erwann O. Michel-Kerjan and Paul A. Raschky, *Does Federal Disaster Assistance Crowd Out Private Demand for Insurance?* (August 2014), available at http://opim.wharton.upenn.edu/risk/library/WP201404_CK-EMK-PAR_Does-assistance-crowd-out-insurance.pdf.

[244] Paul A. Raschky, Reimund Schwarze, Manijeh Schwindt and Ferdinand Zahn, *Uncertainty of Governmental Relief and the Crowding out of Flood Insurance*, 54 Environmental and Resource Economics 179 (2013).

statistically significant negative impact on average coverage per policy. Kousky et al. state, "A $1,000 increase in the average IA [Individual Assistance program] grant decreases average insurance coverage by roughly $6,400".[245] However, the volume of Small Business Administration disaster loans has no significant effect.[246] In other words, government loans induce less crowding-out than direct grants.[247] Counterpart aid under the Whole-Nation System operates much like disaster grants from the IA program and thus would create substantial crowding-out of market-based insurance. Government loans, however, even at a low interest rate, will induce less crowding-out of insurance than direct disaster grants. Therefore, the form of counterpart aid could be changed into government loans rather than direct disaster grants to reduce this crowding-out effect.

V. CONCLUSION

The government is playing an expanding role in catastrophe disaster aid, relief, and compensation all around the world. Government expenditures are sharply rising, not only in the United States and European countries, but elsewhere.[248] The same is happening in China, where government traditionally plays a fundamental role in dealing with catastrophe disasters.

Based on the above analysis, we may admit that the Whole-Nation System is necessary because it works well in the short run as emergency relief, especially in the absence of a catastrophe insurance market. In addition, the government continues this system because it helps the

---

[245] Carolyn Kousky, Erwann O. Michel-Kerjan and Paul A. Raschky, *Does Federal Disaster Assistance Crowd Out Private Demand for Insurance?* (August 2014), available at http://opim.wharton.upenn.edu/risk/library/WP201404_CK-E MK-PAR_Does-assistance-crowd-out-insurance.pdf.

[246] Carolyn Kousky, Erwann O. Michel-Kerjan and Paul A. Raschky, *Does Federal Disaster Assistance Crowd Out Private Demand for Insurance?* (August 2014), available at http://opim.wharton.upenn.edu/risk/library/WP201404_CK-EMK-PA R_Does-assistance-crowd-out-insurance.pdf.

[247] Carolyn Kousky, Erwann O. Michel-Kerjan and Paul A. Raschky, *Does Federal Disaster Assistance Crowd Out Private Demand for Insurance?* (August 2014), available at http://opim.wharton.upenn.edu/risk/library/WP201404_ CK-EMK-PAR_Does-assistance-crowd-out-insurance.pdf.

[248] J.D. Cummins, M. Suher and G. Zanjani, *Federal Financial Exposure to Natural Catastrophe Risk* in Measuring and Managing Federal Financial Risk 61 (D. Lucas ed., 2010); European Commission, *Disaster Risk Reduction: Increasing Resilience by Reducing Disaster Risk in Humanitarian Action* (2013), DG ECHO Thematic Policy Documents, available at http://ec.europa.eu/echo/files/policies/ prevention_preparedness/DRR_thematic_policy_doc.pdf.

government gain support from the people.[249] When disaster provides a unique opportunity for the government to show its responsibility and accountability, it is not difficult to imagine that the Whole-Nation System, with its powerful capability, increases people's support for the government. If only for political expediency, it is hard to believe that the Chinese government will give up the Whole-Nation System completely, even if the system has a lot of problems.

Nonetheless, the Whole-Nation System needs reform. It should be combined with private catastrophe insurance to achieve the benefits of both private market and public government. The Whole-Nation System mainly works well in emergency relief. Beyond that, the government should encourage catastrophe insurance, provide risk finance and loss compensation and support insurance in that role.

Alford once raised the question "whether developing countries have a fateful choice: to embrace Western models of professional organization as they now exist, or to set off on an independent path, adapting elements of Western practices to their own historical and cultural situation".[250] Though the answer may be "blowing in the wind", I wish that the proposed catastrophe insurance market-enhancing framework could help capture some part of that answer. Hopefully, it could shed light on solving the universal dilemma of how to manage catastrophe risk efficiently and cover disaster losses fairly.

---

[249]   The impact of natural disasters on support for authorities is conditional on governmental performance during and after the disaster. For a transitional state like China, government's legitimacy depends not only on its economic performance but also on its response and accountability to the people. A similar example presents in Russia. In the summer of 2010 central Russia experienced unprecedented forest fires. The wildfires were the most disastrous in recorded history in Russia. The fires burned more than 500,000 hectares of land, many people died and over 1,200 houses were destroyed. An empirical study conducted in the areas affected by the disaster over the course of the summer of 2011 revealed that rapid government action and generous aid relief led to increased loyalty to the government among those affected by the disaster. See Y.A. Lazarev, A.S. Sobolev, I.V. Soboleva and B. Sokolov, *Trial by Fire: A Natural Disaster's Impact on Support for the Authorities in Rural Russia*, 66 World Politics 641 (2014).

[250]   William P. Alford and Kenneth Winston, *Introduction* in Prospects for the Professions in China 2 (William P. Alford, William Kirby and Kenneth Winston eds., 2011).

2. Climate change, and financial instruments to cover disasters: what role for insurance in transitional China?

I. INTRODUCTION

Due to climate change and an increasing concentration of the world's population in vulnerable areas, natural catastrophe disasters are predicted to become more frequent, more intense, and more costly in the coming years, with losses increasing dramatically during the past few years.[1] In theory, private insurance can be an efficient financial instrument to cover disasters, as will be explained in section II below. In practice, private insurance plays an important role in developed countries such as the United States. Following from the preceding discussion in Chapter 1 on catastrophe risk management regime, this chapter further addresses these questions: taking into account China's transition economy and specific socialist system, what is the role of private insurance, which is the alternative choice of the Whole-Nation System, to cover disasters, and how does it distribute catastrophe risks sustainably?

II. THE ROLE OF INSURANCE IN CATASTROPHE COVERAGE

Traditionally, the role of the insurance industry has been to distribute risks. The risk management process is based on three pillars: risk assessment (or risk analysis), risk control, and risk financing. Among

---

[1]   Muthukumara Mani, Michael Keen and Paul K. Freeman, *Dealing with Increased Risk of Natural Disasters: Challenges and Options*, IMF Working Paper (2003), available at http://www.imf.org/external/pubs/ft/wp/2003/wp03197.pdf.

these, insurance is an important method of risk financing to transfer risk.[2]

Risk and uncertainty are closely connected concepts, but they should be properly separated from each other. Almost one hundred years ago, Frank Knight, in his book *Risk, Uncertainty and Profit*, carefully distinguished between risk and uncertainty. Risk refers to a *measurable* exposure and uncertainty refers to an *un-measurable* exposure. The uncertainty of an exposure cannot be quantified or put into a numerical value. This difference is relevant to insuring major catastrophes, because when there is "too much" unpredictability, there is too much uncertainty to quantify an exposure to loss.[3] Section III below describes the challenges of dealing with uncertainty in catastrophe management.

Many law and economics scholars favor insurance as a private-market mechanism for distributing catastrophe risk, especially compared to government-provided compensation.[4] For example, Kunreuther,[5] Epstein,[6] Priest,[7] and Kaplow[8] argue that insurance is better equipped to deal with catastrophe risks due to its advantages of lower transaction costs, lower adverse selection, and more efficiency as a result of competitive markets.

A.   The Use of Insurance to Cover Disasters

Insurance is regarded as an important tool to cover losses caused by disasters due to its ability to transfer and reduce risk. Emmett and Therese Vaughan give definitions of insurance from two different viewpoints. From the viewpoint of individuals, insurance is a device to transfer risk, whereas from the viewpoint of society, insurance is a device to reduce

---

[2]   François Outreville, Theory and Practice of Insurance 45–64 (1998). See also Rob Thoyts, Insurance Theory and Practice 286–295 (2010).

[3]   Frank Knight, Risk, Uncertainty and Profit 233 (1971). This book was written in 1921.

[4]   Michael Faure and Klaus Heine, *Insurance against Financial Crises?*, 8 New York University of Business and Law 117 (2011).

[5]   Howard Kunreuther, *The Case for Comprehensive Disaster Insurance*, 11 The Journal of Law and Economics 133, 161–163 (1968).

[6]   Richard A. Epstein, *Catastrophe Responses to Catastrophe Risks*, 12 Journal of Risk and Uncertainty 287, 287 (1996).

[7]   George L. Priest, *The Government, the Market, and the Problem of Catastrophe Loss*, 12 Journal of Risk and Uncertainty 219, 219 (1996).

[8]   Louis Kaplow, *Incentives and Government Relief for Risk*, 4 Journal of Risk and Uncertainty 167, 167 (1991).

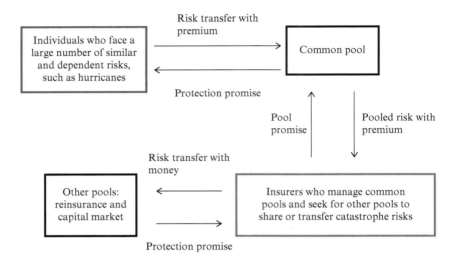

Figure 2.1   The mechanism of insurance to cover disasters

risks through pooling.[9] "Risk transfer" refers to the process by which an individual substitutes a small certain cost, called the premium, for a large uncertain financial loss, which is the contingency insured against; meanwhile, "risk pooling" is the process of combining a sufficient number of homogeneous exposures into a group to make the losses statistically predictable for the group as a whole.[10] Figure 2.1 depicts the process of insuring disasters via risk transfer and risk pooling.

B.   Risk Aversion

In insurance economics, "risk aversion" is commonly used to describe individuals' "attitude to risk".[11] A person is said to be "risk averse" "if she considers the utility of a certain prospect of money income to be higher than the expected utility of an uncertain prospect of equal expected monetary value".[12]

The concept of risk aversion can explain the use of insurance to cover

---

[9]   Emmett J. Vaughan and Therese M. Vaughan, Fundamentals of Risk and Insurance 34–44 (2007).

[10]   Emmett J. Vaughan and Therese M. Vaughan, Fundamentals of Risk and Insurance 34–44 (2007).

[11]   S. Hun Seog, The Economics of Risk and Insurance 18–33 (2010).

[12]   Robert Cooter and Thomas Ulen, Law & Economics 50 (2008).

disasters. People who are risk averse want to transfer risk. Then the question is who would be willing to assume this transferred risk. When a risk seems relatively small (which depends on different individuals' attitudes), a risk-averse individual may become risk neutral. The relatively risk-neutral person may play the role of insurer by taking others' risk. This mechanism is efficient if no one suffers damage and someone becomes happier by transferring risk, satisfying the conditions for "Pareto optimality".[13]

Assuming, however, that all individuals are strictly risk averse, then no one will be interested in taking someone else's risk. Under such circumstances, a possible alternative to risk transfer is risk pooling. In risk pooling, each individual relies on the pool and all members of the pool become relatively risk neutral when they face a larger risk.[14] In other words, the participants transfer their risks to the pool and all contributed premiums to pay for claims for losses on the risks insured in the pool. If individuals' risks are independent of each other, this risk pooling mechanism also conforms to the law of large numbers by which the future risk is more certain in larger groups. This may make the risk easier to underwrite in the aggregate than for any particular individual because it reduces uncertainty.[15]

C.   The Current State of the Catastrophe Insurance Market

Over the last few decades, the number of natural catastrophes has risen.[16] During this time, insurers who underwrote these exposures faced more insured losses. In the United States, Hurricane Andrew struck Florida and

---

[13]   "Pareto optimality", also called "Pareto efficiency", is a state of allocation of resources in which it is impossible to make any one individual better off without making at least one individual worse off. See S. Hun Seog, The Economics of Risk and Insurance 35–39 (2010).

[14]   Karl Borch, *Equilibrium in a Reinsurance Market*, 30 Econometrica 424 (1962).

[15]   The law of large numbers is the mathematical theorem that says that for a series of independent and identically distributed random variables, the variance of the average amount of a claim payment decreases as the number of claims increases. See Howard Kunreuther and Richard J. Roth, Sr, Paying the Price: The Status and Role of Insurance against Natural Disasters in the United States 24–26 (1998). See also Tom Baker, Insurance Law and Policy: Cases and Materials 3 (2008); Véronique Bruggeman, Compensating Catastrophe Victims: A Comparative Law and Economics Approach 59–60 (2010).

[16]   Swiss Re Sigma, *Natural Catastrophes and Man-made Disasters in 2014: Convective and Winters Storms Generate Most Losses* (2015), available at http://www.actuarialpost.co.uk/downloads/cat_1/sigma2_2015_en.pdf.

cost insurers $21.6 billion in 1992 (in 2006 dollars).[17] Hurricane Katrina in 2005 caused about $48 billion in catastrophe-related insured losses—a historic record in the United States.[18] In 2009, Swiss Re and the Insurance Information Institute issued a report titled "Twenty-Five Most Costly Insured Catastrophes Worldwide, 1970–2008". Almost half of the insured catastrophes for the 38 years studied occurred in the previous five years, from 2003 to 2008. The details of the insured catastrophes from 2003 to 2008 are shown in Table 2.1.

The total cost of catastrophes in 2012 was $186 billion worldwide, of which $77 billion was covered by insurers.[19] Therefore, the gap between insured and noninsured losses was $109 billion.[20] Even in the United States, the biggest private insurance market in the world, there are large and growing gaps in catastrophe insurance. After Hurricane Katrina in 2005, numerous insurance firms cut back their coverage in coastal areas. In 2007, Florida responded to noninsured catastrophe risks by setting up its own government-backed insurance company, the Citizens Property Insurance Corporation. As a result of the subsidized price of coverage, this company had become the largest catastrophe insurance underwriter by 2010.[21] In the United States, the free market is regarded as an integral part of the US political economy.[22] However, as insurers have withdrawn from the catastrophe insurance market, the government has intervened. The next section uses a supply-demand framework to explain why the catastrophe insurance market displays these anomalies.

---

[17] Kevin McCarty et al., Task Force on Long-Term Solutions for Florida's Hurricane Insurance Market (2006), available at http://www.myfloridacfo.com/hurricaneinsurancetaskforce/TaskforceRS2/draftlts6.pdf.

[18] Howard Kunreuther and E.O. Michel-Kerjan, *Market and Government Failure in Insuring and Mitigating Natural Catastrophes: How Long-Term Contracts Can Help* in Public Insurance and Private Markets 115 (J.R. Brown ed., 2010).

[19] Swiss Re, Natural Catastrophes and Man-made Disasters in 2012: A Year of Extreme Weather Events in the US (2013), available at http://www.swissre.com/media/news_releases/nr_20130327_sigma_natcat_2012.html.

[20] Swiss Re, Natural Catastrophes and Man-made Disasters in 2012: A Year of Extreme Weather Events in the US (2013), available at http://www.swissre.com/media/news_releases/nr_20130327_sigma_natcat_2012.html.

[21] Howard Kunreuther, Mark V. Pauly and Stacey McMorrow, Insurance and Behavioral Economics: Improving Decisions in the Most Misunderstood Industry 11–12 (2013).

[22] Katherine Swartz, *Justifying Government as the Backstop in Health Insurance Markets*, 2 Yale Journal of Health Policy, Law, and Ethics 89 (2002).

Table 2.1   *The most costly insured catastrophes worldwide from 2003–2008*

$ BILLION	EVENT	VICTIMS (DEAD OR MISSING)	YEAR	AREA OF PRIMARY DAMAGE
17.6	Hurricane Ike	348	2008	USA, Caribbean, et al.
5.0	Hurricane Gustav	153	2008	USA, Caribbean, et al.
6.3	Winter Storm Kyrill	54	2007	Germany, UK, NL, France
48.1	Hurricane Katrina	1,836	2005	USA, Gulf of Mexico, et al.
13.8	Hurricane Wilma	35	2005	USA, Gulf of Mexico, et al.
11.1	Hurricane Rita	34	2005	USA, Gulf of Mexico, et al.
14.6	Hurricane Ivan	124	2004	USA, Caribbean, et al.
9.1	Hurricane Charley	24	2004	USA, Caribbean, et al.
5.8	Hurricane Frances	38	2004	USA, Bahamas
4.4	Hurricane Jeanne	3,034	2004	USA, Caribbean, et al
4.0	Typhoon Songda	45	2004	Japan, South Korea
3.7	Storms	45	2003	USA

Note: Howard Kunreuther and E. O. Michel-Kerjan, At War with the Weather: Managing Large-scale Risks in a New Era of Catastrophes' (2009). These data are from Swiss Re (2009) and Insurance Information Institute in New York and the dollar amounts are indexed to 2008.

III. A SUPPLY–DEMAND FRAMEWORK ANALYSIS OF CATASTROPHE INSURANCE IN COVERING DISASTERS

From the above analysis, we learn how the insurance mechanism can ideally be efficient through risk transfer and risk pooling. Nevertheless, insurers have done relatively little to address catastrophe risks, and have even cut back coverage after exposure to losses. Flood insurance in flood-prone areas of the United States is an example.[23] This quandary raises the question, which barriers prevent insurers from underwriting catastrophe insurance policies? As it turns out, barriers exist on both the supply side and the demand side.[24] This section explores the supply–demand dynamics.

A. The Supply of Catastrophe Insurance

On the supply side, classical economic theory assumes that insurance companies maximize their long-run expected profits in a competitive insurance market.[25] However, in the real world, insurers' behavior often differs from the classical theory.

1. The insurability of catastrophe risk

In theory, catastrophe risks might make good business for insurers because bearing risks for money is the business of insurers. However, insurers decide whether to cover a catastrophe risk based on whether that risk is insurable. A risk is considered insurable if it satisfies at least two requirements: first, the insurer must have the ability to identify, quantify, and estimate the chances of disasters and the resulting losses; and second, the insurer must have the ability to set and collect appropriate premiums for catastrophe risks.[26]

---

[23] Flood insurance was first offered by private insurers in the late 1890s. However, the losses to insurers were so large that the insurers left the market. In 1968, the Congress created the National Flood Insurance Program (NFIP) as an alternative means to offer coverage subsidized by the federal government: see Howard Kunreuther and Richard J. Roth, Sr, Paying the Price: the Status and Role of Insurance against Natural Disasters in the United States 40 (1998).

[24] Sean B. Hecht, *Climate Change and the Transformation of Risk: Insurance Matters*, 55 UCLA Law Review 1559 (2008).

[25] Howard Kunreuther, Mark V. Pauly and Stacey McMorrow, Insurance and Behavioral Economics: Improving Decisions in the Most Misunderstood Industry 18 (2013).

[26] Howard C. Kunreuther and Erwann O. Michel-Kerjan, *Climate Change, Insurability of Large-Scale Disasters, and the Emerging Liability Challenge* 155

(a)   The predictability condition of insurability Many insurable risks, such as house fires or automobile accidents, occur on a regular basis. It is possible to identify and quantify such risks and to estimate losses of such accidents by using historical data. For catastrophes, however, identifying and quantifying risks is more difficult due to the low probability of these disasters and thus limited historical data. Furthermore, the uncertainty of catastrophes increases the difficulty of estimating the frequency and damages of disasters.[27] These obstacles make it almost impossible for catastrophe risks to satisfy the first requirement of insurability, which is the ability to estimate the probability of the risk and quantify the resulting losses.

(b)   Pricing requirements for insurability Price setting is the other challenge for catastrophe risks. Setting premiums is influenced by both intrinsic factors such as the ambiguity of risks, the degree of correlated risks, and asymmetric information about the risk, plus extrinsic factors such as rate regulation and capital market requirements. This section will be limited to an analysis of intrinsic factors. Extrinsic factors will be examined below.

Insurance pricing theory makes a distinction between risk and uncertainty. Uncertainty refers to un-measurable exposure. The uncertainty of an exposure cannot be quantified or assigned a number value. The value of the un-measurable exposure is "unknowable", therefore making it impossible to calculate a rational charge for taking the exposure. In other words, uncertainty leads to unreliable insurance premiums.

Asymmetric information about the risk often includes moral hazard and adverse selection.[28] Moral hazard occurs when the insured has an informational advantage from better knowing his risk and expected losses. The insured will then have a tendency to take risks, since the insurers will bear most of the loss, and thus the possibility of a loss will increase after the insured purchases the insurance.[29] For example, when an individual

---

University of Pennsylvania Law Review 1797 (2007). See also Howard Kunreuther and Richard J. Roth, Sr, Paying the Price: The Status and Role of Insurance against Natural Disasters in the United States 27–38 (1998).

[27]   J.D. Cummins and C.M. Lewis, *Catastrophe Events, Parameter Uncertainty, and the Breakdown of Implicit Long Term Contracting: The Case of Terrorism Insurance*, 26 Journal of Risk and Insurance 154 (2003).

[28]   Asymmetric information exists where one party has more or better information than the other in the transaction.

[29]   Véronique Bruggeman, Compensating Catastrophe Victims: A Comparative Law and Economics Approach 61–64 (2010).

purchases a hurricane insurance policy, he may have no interest in performing mitigation work on his house, such as making it storm proof. When a hurricane happens, the possibility of his house being damaged thus increases. Adverse selection means that those with the highest risk are most likely to buy the insurance.[30] For example, people who live in flood-prone areas are most likely to purchase flood insurance, while others are not. This will lead to a flood insurance pool that is full of high risks.

The avoidance of correlated risks is a very important prerequisite for setting premiums for a large risk pool. The law of large numbers as the basic principle of insurance depends on independent events so the insurer can spread the risk over a large risk pool.[31] Ideally in insurance, a policy-holder pays a small certain premium to avoid a large future loss. The risks need to be independent in order for a risk pool to work. If the risks are correlated, the pool could not afford to pay all the risks simultaneously. In contrast, a catastrophe may cause thousands of concentrated losses. For example, Hurricane Katrina struck thousands of houses in the city of New Orleans. The damaged properties in New Orleans presented a highly correlated risk, which magnified the risk of the insured pool and made pricing premiums unreliable.

2.   Capacity restrictions on insurers

Due to the highly correlated nature and potential concentration of losses from catastrophes, the capacity of the insurance market may not be sufficient to absorb those losses. The potential losses from catastrophe risks are severe and uncertain. As discussed above, Hurricane Andrew caused about $25 billion in losses in Florida. Insurers paid over $21.6 billion in claims related to Hurricane Andrew. Moreover, annual losses are highly variable and may require a large sum of money to cover high losses in some years. Big losses such as Hurricane Andrew led 11 property-casualty insurance companies to become insolvent.[32] Hurricane Andrew revealed that Florida faced a "capacity gap", which was the difference between the amount of the available insurance industry capital and the demand for catastrophe coverage.[33] As a result, insurers and financial market

---

[30]   Véronique Bruggeman, Compensating Catastrophe Victims: A Comparative Law and Economics Approach 64–66 (2010).

[31]   G.L. Priest, *The Government, The Market and the Problem of Catastrophic Loss*, 12 Journal of Risk and Insurance 219, 221–222 (1996).

[32]   US Government Accountability Office, Catastrophe Insurance Risk: The Role of Risk-linked Securities and Factors Affecting Their Use, GAO-02-941 (2002).

[33]   Rawle O. King, *Hurricanes and Disaster Risk Financing Through Insurance Challenges and Policy Options*, Congressional Research Service, R132825 (2005).

experts came to realize that outside capital was needed to supplement the industry's capacity after Hurricane Andrew.[34]

Hurricane Katrina also showed that the insurance industry may not have sufficient capital to cover every mega-catastrophe. Total damages associated with Hurricane Katrina are expected to exceed $200 billion. The federal government expected to spend over $100 billion for response and recovery efforts associated with that hurricane.[35]

3. Profitability requirements for catastrophe insurance

As mentioned above, classical economic theory[36] asserts that insurers supply insurance to maximize their long-run expected profits. But in some circumstances, even when the risk is insurable and the insurer has the capacity to cover the risk, underwriting some risks may not be economically profitable due to other obstacles.

The first such obstacle is rate regulation. In the United States, insurance is mainly regulated at the state level. The "rate" refers to the premium of the policy, which is sometimes regulated. Because of consumer-protection concerns, insurance commissioners may require catastrophe insurance policy premiums to be "affordable", which may prevent insurers from pricing policies to accurately reflect risk.[37] Insurers who have to supply catastrophe insurance at such affordable prices may not be able to make profits.

The second obstacle is the short-run profit horizon of insurers. Many insurance companies are publicly-owned corporations in which there is a separation of ownership and control. Even if the owners/investors are risk neutral and prefer to underwrite a catastrophe risk to maximize long-run expected profits, managers may follow a safety-first rule as a result of risk aversion and fail to underwrite that risk.[38] According to a McKinsey &

---

[34] Rawle O. King, *Hurricanes and Disaster Risk Financing Through Insurance Challenges and Policy Options*, Congressional Research Service, R132825 (2005).

[35] Rawle O. King, *Hurricanes and Disaster Risk Financing Through Insurance Challenges and Policy Options*, Congressional Research Service, R132825 (2005).

[36] Under classical economics, "all human behavior can be viewed as involving participants who (1) maximize their utility (2) from a stable set of preferences and (3) accumulate an optimal amount of information and other inputs in a variety of markets". See Véronique Bruggeman, Compensating Catastrophe Victims: A Comparative Law and Economics Approach (2010). In other words, the classical economics literature starts from the assumption that human beings make their decisions based upon rational analysis, such as an objective weighing of costs and benefits.

[37] Sean B. Hecht, *Climate Change and the Transformation of Risk: Insurance Matters*, 55 UCLA Law Review 1559 (2008).

[38] The safety-first rule (or safety-first model) is a model of insurer pricing that reflects the insurer's threshold probability that losses for a specific event will not

Company report from 2005, "shortsighted behavior is widespread" among managers, and their myopic behavior shows their disfavor to developing products for long-run profits.[39] Insurance company managers are not immune to this myopia.

The third obstacle is the "appetite" of insurers.[40] If insurers have no appetite for it, they will decline to write policies. The appetite of insurers is mainly affected by their estimation on the insurability of catastrophe risk. As noted by Trevor Maynard of Lloyd's, insurers "cannot insure our way out of the [climate change] problem" because "[r]einsurers and alternative capital market providers will not accept risk on terms that are not commercially viable".[41]

4.   A short summary of the supply of catastrophe insurance
From the above discussion, we can see that insurers face a number of obstacles in underwriting catastrophe insurance policies. The nature of catastrophe risks makes it difficult to fulfill all of the insurability requirements. Due to the high potential losses from catastrophe exposures, insurers' capacity and appetite to cover such losses is not sufficient. As a result, the supply of catastrophe insurance is limited and volatile.

B.   The Demand for Catastrophe Insurance

On the demand side, classical economic theory posits that individuals will make decisions under uncertainty according to the "expected utility theory of choice".[42] According to Nobel Prize winner Kenneth J. Arrow,

---

exceed a pre-specified value. See Howard Kunreuther, Mark V. Pauly and Stacey McMorrow, Insurance and Behavioral Economics: Improving Decisions in the Most Misunderstood Industry 146–154 (2013).

[39]   Richard Dobbs, Keith Leslie and Lenny T. Mendonca, *Building the Healthy Corporation*, McKinsey Quarterly 63, 65 (2005), available at https://www.mck insey.com/business-functions/organization/our-insights/building-the-healthy-corpo ration.

[40]   Appetite is a subjective criterion of profitable restriction, which generally refers to the willingness of insurers to underwrite policies. Inspired by Lecturer Douglas Simpson.

[41]   Trevor Maynard, *Climate Change: Impacts on Insurers and How They Can Help With Adaptation and Mitigation*, 33 The Geneva Papers on Risk and Insurance 140 (2008).

[42]   This theory assumes that individuals with accurate information about risks decide on insurance purchases by making explicit tradeoffs between the expected benefits and the costs of different policies. See Howard Kunreuther, Mark V. Pauly and Stacey McMorrow, Insurance and Behavioral Economics: Improving Decisions in the Most Misunderstood Industry 8 (2013).

individuals purchase insurance because they are willing to pay a certain small premium to protect against an uncertain large loss.[43] In other words, a rational potential victim residing in a hazard-prone area will voluntarily purchase catastrophe insurance if he perceives the premium to be sufficiently low in comparison to the risks.

However, many people fail to purchase insurance offered even at subsidized prices against low-probability but high-consequence disasters.[44] Flood insurance offered by the National Flood Insurance Program, whose rates and terms are set by Congress, is an example of this anomaly. During the program's first four years, fewer than 3,000 of 21,000 flood-prone communities entered into the subsidized flood insurance program; by 1992, a conservative estimate of coverage suggested that less than 20 percent of homeowners located in the flood-prone areas bought flood insurance.[45] In 1997, the Federal Insurance Administration estimated that only about 27 percent of households living in high-risk flood areas were insured.[46] There are several theories as to why demand for catastrophe insurance coverage is so low, as follows.

1. Intuitive thinking versus deliberative thinking

In Daniel Kahneman's book *Thinking: Fast and Slow*, the author adopted terms of thinking originally proposed by the psychologists Keith Stanovich and Richard West, which he labeled System 1 and System 2:

> *System 1* operates automatically and quickly, with little or no effort and no sense of voluntary control. This is often described as intuitive thinking.
>
> *System 2* allocates attention to effortful and intentional mental activities including simple or complex computations or formal logic. The operations are often associated with the subjective experience of agency, choice, and concentration. This is often described as deliberative thinking.[47]

---

[43]   Kenneth J. Arrow, Essays in the Theory of Risk-Bearing 199–200 (1971).

[44]   Howard Kunreuther, Mark V. Pauly and Stacey McMorrow, Insurance and Behavioral Economics: Improving Decisions in the Most Misunderstood Industry 113 (2013).

[45]   Howard Kunreuther and Richard J. Roth, Sr, Paying the Price: The Status and Role of Insurance against Natural Disasters in the United States 55 (1998).

[46]   Howard Kunreuther and Richard J. Roth, Sr, Paying the Price: The Status and Role of Insurance against Natural Disasters in the United States 55 (1998).

[47]   Daniel Kahneman, Thinking, Fast and Slow 20–21 (2011).

When people are operating under System 2, they make choices and decide what to think about and what to do through conscious thought and reasoning.[48] If, as classical economic theory posits, consumer behavior will follow expected utility theory, then consumers (the decision makers) will follow System 2 to make deliberative choices. In reality, much human behavior conforms to the more automatic and less analytic System 1, which results in many biases and simplified anomalies. When consumers face catastrophic disasters, however, System 1 (intuitive thinking) does not work well and results in many anomalies in the demand for catastrophe insurance, such as consumers' failure to buy insurance optimally.

2. Prospect theory

Prospect theory, which was developed by Daniel Kahneman and Amos Tversky, is a descriptive choice model that predicts actual behavior better than expected utility theory.[49] This model is helpful in explaining consumer anomalies in purchasing catastrophe insurance.

The value function of prospect theory that separates losses from gains, states that the pain from a certain loss will be viewed as larger than the positive feeling from an uncertain gain.[50] In other words, people tend to be loss averse. For example, when consumers confront a 20 percent chance of losing $100 versus the certainty of losing $20, they will avoid the certain loss of $20 and take the risk of losing $100. According to the value function of prospect theory as applied to insurance, consumers are more willing to take an uncertain risk than suffer a certain loss in the form of a premium payment.[51] This tendency to treat a certain loss as more painful than the pleasure of uncertain gains is also termed "myopic loss

---

[48] Daniel Kahneman, Thinking, Fast and Slow 15 (2011).

[49] Daniel Kahneman and Amos Tversky, *Prospect Theory: An Analysis of Decision under Risk*, 47 Econometrica 263 (1979).

[50] Value function is a term borrowed from mathematics. As an illustration, imagine the coordinate system, in which the x-axis depicts the magnitude of the gain or loss, and the y-axis, which is y(x) and y(-x), represents the value associated with a gain (x) or loss (-x). The value function is steeper in the loss domain than in the gain domain. It shows in the coordinate system that the curve of y(-x) is steeper than the curve of y(x). See Howard Kunreuther, Mark V. Pauly and Stacey McMorrow, Insurance and Behavioral Economics: Improving Decisions in the Most Misunderstood Industry 96–98 (2013).

[51] Howard Kunreuther, Mark V. Pauly and Stacey McMorrow, Insurance and Behavioral Economics: Improving Decisions in the Most Misunderstood Industry 115 (2013).

aversion".[52] It makes even actuarially fair insurance unattractive, let alone low-probability and high-consequences catastrophe insurance, where it is difficult to charge accurate premiums.

As a result of loss aversion, consumers tend to choose not to buy catastrophe insurance. For extreme catastrophe risks, consumers will just ignore the risk. Prior to a disaster, they contend it will not happen to them. As a result, they will not spend money to invest in protective measures, such as catastrophe insurance.[53] However, when the disaster truly happens, they may feel remorse that they did not buy insurance.

3. The goal-based model of choice

The goal-based model of choice developed by David H. Krantz and Howard Kunreuther is another theory of decision making based on preset goals rather than on maximizing expected utility.[54] In the area of insurance, these goals are categorized as (1) sharing risk for financial protection, (2) earning returns on investments, (3) satisfying emotion-related goals, (4) fulfilling legal or other official requirements, and (5) satisfying social and/or cognitive norms.[55]

The goal-based model also shows that consumers do not always maximize their expected utility.[56] Flood insurance is an example. Before a flood strikes, residents seldom buy flood insurance to protect themselves. At this stage, there are concerns that buying flooding insurance is not a good investment in view of the anticipated benefits. But after suffering flood damage, they purchase insurance to satisfy emotional goals. Following flood damage, anxiety is high, and reducing it by purchasing insurance is a salient goal.[57] However, when several consecutive years pass with no flood, many people cancel their flood policies.[58] At this stage, avoiding

---

[52] Shlomo Benartzi and Richard Thaleer, *Myopic Loss Aversion and the Equity Premium Puzzle*, 110 Quarterly Journal of Economics 73 (1995).

[53] H. Kunreuther and E.O. Michel-Kerjan, *Market and Government Failure in Insuring and Mitigating Natural Catastrophes: How Long-Term Contracts Can Help* in Public Insurance and Private Markets 126–130 (J.R. Brown ed., 2010).

[54] David H. Krantz and Howard Kunreuther, *Goals and Plans in Decision Making*, 2 Judgment and Decision Making 137 (2007).

[55] David H. Krantz and Howard Kunreuther, *Goals and Plans in Decision Making*, 2 Judgment and Decision Making 137 (2007).

[56] Sean B. Hecht, *Climate Change and the Transformation of Risk: Insurance Matters*, 55 UCLA Law Review 1559 (2008).

[57] Howard Kunreuther, Mark V. Pauly and Stacey McMorrow, Insurance and Behavioral Economics: Improving Decisions in the Most Misunderstood Industry 103 (2013).

[58] Erwann Michel-Kerjan, Sabine Lemoyne de Forges and Howard

anxiety is not that important, and reducing burdensome premiums becomes more important.

4.   Relying on a government bailout

Another reason why some people do not buy catastrophe insurance is that they believe that if they suffer catastrophic damages the government should and will bail them out.[59] Due to government compensation, individual incentives to buy insurance are diminishing. For example, in the United States, the federal government normally provides public disaster relief if a state declares an emergency following a natural disaster.[60] If the government's response falls short after natural disasters, the government faces heavy political pressure. Criticism of the response of former President Bush's administration to Hurricane Katrina is an example.[61] This reason for people's reduced demand for insurance is not really anomalous but rather based on rational behavior.

However, relying on government disaster relief leads to problems. A government bailout may lead to a "natural disaster syndrome" in which people fail to voluntarily adopt cost-effective loss-reduction measures.[62] This will lead to the "Samaritan's Dilemma", in which providing relief will further reduce residents' incentives to invest in protective measures such as buying insurance and mitigation.[63]

5.   A short summary of the demand for catastrophe insurance

Consumer behavior related to purchasing catastrophe insurance deviates from classical economic theory. Scholars have used models from behavioral economics to explain such anomalies. These models include intuitive

---

Kunreuther, *Policy Tenure under the U.S. National Flood Insurance Program*, 32 Risk Analysis 644 (2012).

[59]   Michael Faure and Véronique Bruggeman, *Catastrophic Risks and First-Party Insurance*, 15 Connecticut Insurance Law Journal 1 (2008); Sean B. Hecht, *Climate Change and the Transformation of Risk: Insurance Matters*, 55 UCLA Law Review 1559 (2008); Howard Kunreuther, Mark V. Pauly and Stacey McMorrow, Insurance and Behavioral Economics: Improving Decisions in the Most Misunderstood Industry 114–115 (2013).

[60]   Howard Kunreuther, Mark V. Pauly and Stacey McMorrow, Insurance and Behavioral Economics: Improving Decisions in the Most Misunderstood Industry 114–115 (2013).

[61]   "Katrinagate" fury spreads to US media (2005), available at: http://tvnz.co.nz/view/page/425822/609550.

[62]   Howard Kunreuther, *Mitigating Disaster Losses through Insurance*, 12 Journal of Risk and Uncertainty 171 (1996).

[63]   Michael Faure and Véronique Bruggeman, *Catastrophic Risks and First-Party Insurance*, 15 Connecticut Insurance Law Journal 1 (2008).

and deliberative thinking, prospect theory, and the goal-based model of choice. Additionally, due to repeated government bailouts, consumers' choice not to buy catastrophe insurance looks rational.

C. A Short Conclusion: The Feasibility of Catastrophe Insurance

From the above supply-demand framework analysis, it is clear that underwriting catastrophe insurance faces both supply-side and demand-side barriers. On the supply side, problems with insurability and capacity hamper the underwriting process. Meanwhile, on the demand side, consumers buy insufficient catastrophe insurance as a result of behavioral anomalies.

The next section will apply this supply-demand framework to analyze China's catastrophe insurance market and explain why no catastrophe insurance system has been instituted even since the Great Sichuan Earthquake of 2008.[64] Subsequently, some solutions are proposed for the feasibility of catastrophe insurance in China in order to optimize the role of private insurance in covering disasters.

IV. CATASTROPHE INSURANCE IN CHINA: HOW DOES INSURANCE PLAY A ROLE IN COVERING DISASTERS?

A. Why Catastrophe Insurance has Not been Quickly Established since the Great Sichuan Earthquake

In this section, the supply-demand framework presented in section III above is applied to the Chinese context to offer an explanation for why catastrophe insurance was not quickly established after the Great Sichuan Earthquake. To date, some major catastrophic exposures, such as earthquake perils, have not been included in homeowners' insurance.

---

[64] In some countries, for example Turkey and New Zealand, catastrophe insurance systems have been established one or two years after a catastrophe strikes. However, five years after the 2008 earthquake, catastrophe insurance has still not been established in China. See Y.L. Zhou (Vice President of CIRC), *Quan Shehui Diyu Fengxian Nengli Buduan Tigao [The Ability of Defensing Catastrophe Risk is Improving for the Whole Society]*, International Catastrophe Insurance Fund Management Symposium, (2008), available at http://insurance. hexun.com/2008-10-23/110361174.html. See CIRC Is Promoting Regulations of Catastrophe Insurance, and Two Places Are on the Experiment Project, available at http://www.iic.org.cn/D_newsDT/newsDT_read.php?id=106741.

1. Supply restrictions

(a)   Lack of catastrophe data Lack of catastrophe data makes it difficult for insurers to identify, quantify, and estimate the chance of disasters and to set premiums for catastrophe risks. Up to now the acquisition of such data has not been completed in China and is ongoing. For example, from 1998 to 2002 the Swiss Reinsurance Company cooperated with Beijing Normal University to complete the "Digital Map of China Catastrophe Events" (*Zhongguo Juxing Dianzi Zainan Ditu*), which includes historical data, geographic data, weather data, and other types of data from the twelfth century.[65] This digital map will be very helpful in the pricing of catastrophe insurance. In 2009, the China Insurance Regulation Commission (CIRC) promulgated the Catastrophe Insurance Data Acquisition Regulation (JR/T0054-2009), which specified the standard for the acquisition of catastrophe insurance data.[66] Meanwhile, the "Catastrophe Risk Evaluating and Standardizing Target Accumulations" (CRESTA) is developing a new catastrophe risk division system for China.[67] This new system demarcates the existing 60 partitions into 2,472 partitions, which substantially improves the data's accuracy and transparency.[68] However, this project is still ongoing.[69]

(b)   The relatively low capacity of China's insurance industry Along with the rapid growth of China's economy, the growth of the Chinese insurance industry has been impressive. However, China's primary insurance industry still has much less capacity to deal with catastrophe risks than

---

[65]   X. Guo and X. Wei, *The Difficulties and Solutions for Issuing Catastrophe bonds in China [Woguo Faxing Juzai Zhaiquan de Nandian yu Duice]*, 6 China Insurance 23 (2005).

[66]   CIRC Order 52 (2009).

[67]   The CRESTA organization was established by the insurance and reinsurance industry in 1977 as an independent body for the technical management of natural hazard coverage. CRESTA's main goal is to establish a uniform and global system to transfer, electronically, aggregated exposure data for accumulation risk control and modeling among insurers and reinsurers. Available at https://www.cresta.org.

[68]   Swiss Re Beijing Branch, available at http://eol.yzu.edu.cn/eol/common/script/onlinepreview.jsp?countadd=1&lid=2912&resid=234015.

[69]   L. Tian and J. Luo, *The Choice of China Catastrophe Insurance Institution Based on the Restrictions of Supply and Demand—The Feasibility Study of Long Term Catastrophe Insurance [Gong Xu Shuang Yue Shu Xia Zhong Guo Ju Zai Bao Xian Zhi Du De Xuan Ze—Chang Qi Ju Zai Bao Xian De Ke Xing Xing Yan Jiu]*, 65 Wuhan University Journal (Philosophy & Social Sciences) [Wuhan Daxue Xuebao (Zhexue Shehui Kexue Ban)] 111 (2012).

Table 2.2   Insurance penetration and density of certain countries

	China	US	Japan	Netherlands
Insurance penetration: premiums as a % of GDP in 2012	2.96	8.18	11.44	12.99
Insurance density: premiums per capita in USD in 2012	178.9	4047.3	5167.5	5984.9

Source: Swiss Re, World Insurance in 2012 Progressing on the Long and Winding Road to Recovery, (2013) available at http://media.swissre.com/documents/sigma3_2013_en.pdf.

Table 2.3   Comparison between the capital (total reserves) of main property insurers and natural disaster losses in China (RMB billions)

	2007	2008	2009	2010
Net Capital of main insurers	33.8	31.45	43.04	55.88
Natural Disaster losses	236.3	1175.2	252.3	533.99

Source: The figures are from the volumes of Yearbook of China Insurance (2008–2011).

leading Western companies, although it has been growing fast since 1979. The data presented in Table 2.2 show China's lower insurance penetration and insurance density compared to advanced markets.

Similarly, the total capital of China's property insurance companies is much lower than the total amount of losses caused by natural disasters. Table 2.3 shows the existence of this big gap.

Reinsurance is an important potential complement for expanding the capacity of underwriting risks. However, reinsurance currently does not supply strong support for catastrophe insurance in China. At present, the China Reinsurance [Group] Corporation (China Re) is the only state-owned reinsurance group in China, with capital of $6.068 billion. China Re is in a monopoly position and occupies about 80 percent of the aggregate reinsurance market in China. Although China's reinsurance market became open to foreign reinsurance companies after China's entry into the World Trade Organization, only a few reinsurance companies, such as Swiss Re and Munich Re, have established business operations in China, and they are only in the early stages of reinsuring risks.

2. Demand restrictions

(a)   Irrational behavior by consumers Insurance had a late start in China. In 1979, the central government announced the "Notice on Restoration of the Domestic Insurance Business and Strengthening of the Insurance Agency" (*Guanyu Huifu Guonei Baoxian Yewu He Jiaqiang Baoxian Jigou De Tongzhi*).[70] As a result, private insurance agencies resumed their businesses in China.[71] Yet, the insurance market and products are not well-developed. Chinese consumers have weak incentives to purchase insurance products. Because of their low-probability nature, consumers' awareness of and demand for insurance for catastrophe disasters are even weaker.[72] Consumers do not always behave rationally and maximize their expected utility to protect themselves from catastrophe losses by buying insurance. Myopic loss aversion can also explain this anomalous behavior. Prior to a disaster, consumers believe that natural disasters will not happen to them. In addition, they regard the premium as a certain loss, which is more painful than the possible future gains.

Before the Great Sichuan Earthquake, few people and enterprises bought earthquake insurance even though they lived in earthquake-prone areas. However, after that earthquake's severe destruction, the sale of property insurance products including earthquake insurance increased dramatically.[73] According to the goal-based model of choice, avoiding

---

[70]   Available at: http://www.china.com.cn/aboutchina/txt/2009-11/25/content_18951984.htm.

[71]   On October 20, 1949, the People's Insurance Company of China (PICC) was established. However, in the socialist planned economy framework, there was no need for business insurance. From 1958 to 1978, the insurance business virtually disappeared. It was not until 1979, after the 3rd Plenary Session of the 11th Chinese Communist Party (CPC) Central Committee and the process of institutional transformation towards a market economy began, that the State Council gave its approval for the Conference of Branch Heads of the People's Bank of China (PBC) to gradually revive domestic insurance operation.

[72]   T. Yue, J.H. Ning and M.Y. Fan, *The Research on the Establishment of Chinese Catastrophe Insurance and Reinsurance System [Wo Guo Ju Zai Bao Xian Zai Bao Xian Ti Xi De Gou Jian Yan Jiu]*, 16 Foreign Investment in China [ZhongGuo Waizi] 255 (2013).

[73]   L. Tian and J. Luo, *The Choice of China Catastrophe Insurance Institution Based on the Restrictions of Supply and Demand—The Feasibility Study of Long Term Catastrophe Insurance [Gong Xu Shuang Yue Shu Xia Zhong Guo Ju Zai Bao Xian Zhi Du De Xuan Ze—Chang Qi Ju Zai Bao Xian De Ke Xing Xing Yan Jiu]*, 65 Wuhan University Journal (Philosophy & Social Sciences) [Wuhan Daxue Xuebao (Zhexue Shehui Kexue Ban)] 111 (2012).

anxiety over earthquakes has become an overriding goal following the Great Sichuan Earthquake.

(b) *Relying on a government bailout* Consumer demand for catastrophe insurance in China has also been blunted by expectations of a government bailout. Historically speaking, China is a country where the government paid a lot of attention to preventing and distributing catastrophe losses. In the Qing Dynasty, for example, "Records of Laws and Systems of Qing Dynasty" (*Da Qing Hui Dian Shi Li*) listed 12 articles on "Disaster Defense and Reduction Policies".[74] Chinese consumers are even more reliant on government bailouts than consumers in developed market economies such as the United States because of China's recent experience with a state-controlled economy and the lack of a private insurance sector from the early 1950s through 1978. To date, the government still plays a major role in distributing catastrophe risk and compensating victims.[75] Qinghua Xian and Xiaolan Ye use public crisis management theory[76] to analyse the Chinese government's experience in distributing catastrophe risk.[77] After the Great Sichuan Earthquake in 2008, the central and local governments played a key role in combining and allocating the resources of society in disaster relief, and earned a good reputation.[78] Counterpart

---

[74] The measures include the following: food supply, river control and levee building, eradication of locusts, information dissemination and so on. See H. Chen, *Disaster Defense and Reduction Policies in the Qing Dynasty [Qingdai fangzai Jianzai De Zhengce Yue Cuoshi]*, 3 Studies in Qing History [Qingshi Yanjiu] 41 (2004).

[75] Y.L. Zhou (Vice President of CIRC), The Ability of Defensing Catastrophe Risk is Improving for the Whole Society [*Quan Shehui Diyu Fengxian Nengli Buduan Tigao*], International Catastrophe Insurance Fund Management Symposium (2008), available at http://insurance.hexun.com/2008-10-23/110361174.html.

[76] There are four main criteria to measure the performance of public sector crisis management. First, the question is whether or not the government has a crisis management system within its organization. The second criterion is the sensitivity of the government to multi-identities (including ethnicity, class, age, and gender) while rescuing victims. The third criterion is related to the decision-making strategy in a crisis situation. Finally, the fourth criterion is how successfully the government can adjust its bureaucratic norms to emergent norms in the crisis situation. See Fadillah Putra, *Crisis Management in Public Administration*, 13 Planning Forum 152 (2009).

[77] Q.H. Xian and X.L. Ye, *The Practices of Chinese Government Catastrophe Risk Management and Its Role Definition*, 23 Journal of Insurance Professional College 43 (2009).

[78] Z. Qin and L.Q. Chen, *Mode of Managing Catastrophic Risk in China*, 11 Journal of Wuhan University of Science and Technology (Social Science Edition) 33 (2009).

aid between local governments is also a widely used method to distribute catastrophe risk.

Because of their reliance on government disaster relief, however, Chinese citizens have reduced incentives to buy catastrophe insurance. According to an empirical study of catastrophe insurance in five Chinese provinces, there is a negative correlation between government relief and insurers' property and causality premium income.[79] Offering "free" government relief creates a heavy and unpredictable burden on the government's financial budget. Furthermore, it also leads to the "Samaritan's Dilemma", which further reduces consumers' incentives to invest in protection measures such as buying catastrophe insurance and mitigation.

B.   Mandatory Multi-year Insurance as a Possible Solution for Insuring Catastrophes

Faced with the "Samaritan's Dilemma", the Chinese government is encouraging the establishment of a private catastrophe insurance system where insurers play a more important role. After years of deliberation, especially after the Great Sichuan Earthquake in 2008, establishing a catastrophe insurance system became a national priority in China.[80] Since the "Regulation on Agriculture Insurance" (which includes stipulations on catastrophic natural disaster insurance) took effect in March 2013, CIRC has been drafting the "Regulation on Catastrophe Insurance".[81] CIRC has also approved two catastrophe insurance pilot programs, the Shenzhen City Program and the Yunnan Province Program.[82] CIRC's

---

[79]   L. Tian and Y. Zhang, *Influence Factors of Catastrophe Insurance Demand in China—Panel Analysis in a Case of Insurance Premium Income of Five Provinces [Woguo Juzai Baoxian Xuqiu Yingxiang Yinsu Shizheng Yanjiu: Jiyu Wusheng Bufen Baofei Shouru Mianban Yanjiu]*, 26 Wuhan University of Technology (Social Science Edition) [Wuhan Ligong Daxue Xuebao (Shehui Kexue Ban)] 175 (2013).

[80]   Y.L. Zhou (Vice President of CIRC), *Quan Shehui Diyu Fengxian Nengli Buduan Tigao [The Ability of Defensing Catastrophe Risk is improving for the Whole Society]*, International Catastrophe Insurance Fund Management Symposium, (2008), available at http://insurance.hexun.com/2008-10-23/110361174.html.

[81]   Available at http://www.iic.org.cn/D_newsDT/newsDT_read.php?id=106741.

[82]   B. Dong (Vice Director of CIRC Property Insurance Department), *ZhuBu Jianli Fuhe Zhongguo Guoqing De Juzai Baoxian Zhidu [Gradually Establishing Catastrophe Insurance Program which Conforms to China's Specific National Condition]*, speech at 23rd F.A.I.R. conference in September 2013 (2013), available at http://www.sinoins.com/zt/2013-09/18/content_74763.htm.

Shenzhen Bureau will disclose its "Catastrophe Insurance Project in Shenzhen City" in the near future.[83]

As the above analysis shows, the insurance market faces both supply-side and demand-side obstacles to catastrophe coverage. How to increase the use of that insurance is an important question. Howard Kunreuther and other professors have proposed "multiyear insurance contracts" to overcome such obstacles.[84] Kunreuther proposed mandatory insurance several decades ago.[85] Michael Faure and Véronique Bruggeman also suggest that mandatory/compulsory insurance could solve the lack of demand and/or supply.[86] This suggestion is further supported by Anastasia Telesetsky, who treats mandatory catastrophe insurance as a risk-sharing mechanism serving the goals of both corrective and distributive justice.[87]

1. Multiyear insurance

Multiyear insurance refers to insurance in which policies are sold for consecutive years rather than only for one year, and are tied to the

---

[83] *Establishing Catastrophe Insurance System Should Allow Flexibility [Jianli Juzai Baoxian Zhidu Buke Yidaoqie']*, Financial Times (China) (November 13, 2013), available at http://www.zgjrw.com/News/20131113/2013cfn/14423126600.shtml.

[84] Howard Kunreuther, Mark V. Pauly and Stacey McMorrow, Insurance and Behavioral Economics: Improving Decisions in the Most Misunderstood Industry 228–243 (2013); Howard Kunreuther and E.O. Michel-Kerjan, *Market and Government Failure in Insuring and Mitigating Natural Catastrophes: How Long-Term Contracts Can Help* in Public Insurance and Private Markets 115 (J.R. Brown ed., 2010); Howard Kunreuther and E.O. Michel-Kerjan, At War with the Weather: Managing Large-scale Risks in a New Era of Catastrophes 333–350 (2009); Howard Kunreuther and E.O. Michel-Kerjan, *Encouraging Adaptation to Climate Change: Long-Term Flood Insurance*, Resources for the Future, Issue Brief 09-13 (2009), available at http://www.rff.org/files/sharepoint/WorkImages/Download/RFF-IB-09-13.pdf.

[85] Howard Kunreuther, *The Case for Comprehensive Disaster Insurance*, 11 The Journal of Law and Economics 133 (1968).

[86] Michael G. Faure, *Insurability of Damage Caused by Climate Change: A Commentary*, 155 University of Pennsylvania Law Review 1875 (2007); Michael Faure and Véronique Bruggeman, *Catastrophic Risks and First-Party Insurance*, 15 Connecticut Insurance Law Journal 1 (2008); Michael Faure and Véronique Bruggeman, Green Paper on the Insurance of Natural and Man-made Disasters (COM (2013) 213 final): Reaction of the Malta FORUM of Legal Experts on Climate Change Adaptation (2013), available at http://ec.europa.eu/internal_market/consultations/2013/disasters-insurance/docs/contributions/non-registered-organisations/forum-of-european-academic-legal-experts-on-climate-change-adaptation-university-of-malta_en.pdf.

[87] Anastasia Telesetsky, *Insurance as a Mitigation Mechanism: Managing International Greenhouse Gas Emissions through Nationwide Mandatory Climate Change Catastrophe Insurance*, 27 Pace Environmental Law Review 691(2010).

property as opposed to the property owner, as in the traditional annual policy.[88] For example, a policyholder pays a single premium for a five-year policy, but then sells the property and moves away after three years. He is not entitled to a refund of the premium payment for the last two years. Multiyear insurance can satisfy a fundamental objective better than annual policies—providing financial protection against catastrophic losses.[89] Thus, it can be a strong tool to solve some of the supply-side and demand-side challenges to the provision of catastrophe insurance.

On the supply side, multiyear insurance enlarges insurers' capacity by extending the term of an insurance policy. When a catastrophe occurs, multiyear insurance can spread aggregated losses over different years. Moreover, multiyear insurance can reduce transaction costs for both insurers and consumers. For example, consumers' search cost of renewing policies and insurers' marketing cost of underwriting policies can be significantly reduced compared to annual insurance policies.[90]

On the demand side, consumers generally like to purchase insurance after suffering catastrophe damage, such as floods, to reduce anxiety. When several consecutive years pass without any flood, however, many people cancel their flood policies. Multiyear insurance has the potential to deal with this problem, while annual policies do not, because under the annual policies, an insured can cancel the next year's coverage easily. Multiyear insurance would also encourage consumers to invest in cost-effective loss-reduction measures in exchange for premium reduction or another bonus.[91]

However, multiyear insurance can pose some challenges. Consumers who buy a multiyear policy will be more concerned about insurers' financial solvency over a long period.[92] In addition, it does not solve consumers'

---

[88]   Howard Kunreuther, Mark V. Pauly and Stacey McMorrow, Insurance and Behavioral Economics: Improving Decisions in the Most Misunderstood Industry 228–243 (2013); Howard Kunreuther and E.O. Michel-Kerjan, At War with the Weather: Managing Large-scale Risks in a New Era of Catastrophes 333–350 (2009).

[89]   Howard Kunreuther, Mark V. Pauly and Stacey McMorrow, Insurance and Behavioral Economics: Improving Decisions in the Most Misunderstood Industry 228–232(2013).

[90]   Howard Kunreuther, Mark V. Pauly and Stacey McMorrow, Insurance and Behavioral Economics: Improving Decisions in the Most Misunderstood Industry 233 (2013).

[91]   Howard Kunreuther, Mark V. Pauly and Stacey McMorrow, Insurance and Behavioral Economics: Improving Decisions in the Most Misunderstood Industry 233–238 (2013).

[92]   Howard Kunreuther, Mark V. Pauly and Stacey McMorrow, Insurance and Behavioral Economics: Improving Decisions in the Most Misunderstood Industry 236 (2013).

reliance on a government bailout. These problems may be solved by mandating catastrophe insurance, as discussed in the next section.

2. Mandatory insurance

Mandatory insurance can improve efficiency, correct market failure, promote distributive justice, price accurately, and regulate more efficiently.[93] A well-known mandatory insurance example is the French property damage insurance, which requires property owners to pay a supplementary premium for mandatory natural disaster coverage.[94] Mandatory disaster coverage is often discussed as a tool to solve the above supply and demand anomalies.

First, an insurance mandate helps correct irrational behaviors. If individuals do not buy insurance based on the mistaken belief that they are immune to disasters, such an arrangement improves their long-term interests by compelling them to insure.[95]

Second, mandatory coverage helps manage adverse selection. Adverse selection is a significant challenge for the insurance business because those with the highest risk are most likely to buy the insurance while others who are at lower risk are not, and this leads to a pool full of high risks.[96] By forcing universal participation in the risk pool, mandatory insurance prevents lower-risk groups from opting out of the pool.[97] This leads to high insurance penetration, which will enhance the spreading of risk and reduce ad hoc government relief.[98]

For example, in 2000, the Turkish Catastrophe Insurance Pool (TCIP) was established as mandatory insurance, according to the enactment of Decree Law No. 587.[99] The TCIP sold more than 4.8 million policies in 2012, compared to 600,000 policies when the pool was set up, and the

---

[93] Michael G. Faure, *Insurability of Damage Caused by Climate Change: A Commentary*, 155 University of Pennsylvania Law Review 1875 (2007).

[94] Michael Faure and Véronique Bruggeman, *Catastrophic Risks and First-Party Insurance*, 15 Connecticut Insurance Law Journal 1 (2008).

[95] David Moss, When All Else Fails: Government as the Ultimate Risk Manager 51 (2004).

[96] Véronique Bruggeman, Compensating Catastrophe Victims: A Comparative Law and Economics Approach 64–66 (2010).

[97] David Moss, When All Else Fails: Government as the Ultimate Risk Manager 50 (2004).

[98] Youbaraj Paudel, *A Comparative Study of Public–Private Catastrophe Insurance Systems: Lessons from Current Practices*, 37 The Geneva Papers on Risk and Insurance-Issues and Practice 257 (2012).

[99] The decree was enacted on December 27, 1999. However, the article authorizing policy sales did not become effective until a later date due to the need to complete preparatory technical work and adequately inform prospective policyholders in advance about this mandatory insurance. See Eugene Gurenko,

penetration rate rose up to 29 percent nationwide.[100] In contrast, data show that if there is no mandatory purchase requirement, penetration rates are low.[101] In European countries, surveys of the insurance industry have shown that without mandatory insurance coverage, the demand for flood insurance is very low, with a penetration rate of only about 10 percent of private flood insurance coverage in some countries.[102]

Third, compulsory insurance helps enhance damage mitigation. Telesetsky proposes that "the most important reason for mandating catastrophe risk insurance is to compel industry actors to take action under the supervision of the profit-motivated insurance industry".[103] Under a mandate, individuals will want to undertake mitigation investments in order to lower their insurance premiums.[104] As a result, these damage mitigation measures will further lower the aggregate exposure of society, enhance financial solvency, and decrease the costs of the catastrophe insurance system in the long run.[105] In addition, mitigation investments may reduce current incentives to build in high-risk areas.[106]

3.   Mandatory multiyear insurance

From the discussion on multiyear insurance and mandatory insurance, it follows that we may combine these two tools to wed their merits while

---

Earthquake Insurance in Turkey: History of the Turkish Catastrophe Insurance Pool 62 (2006).

[100]   Turkish Catastrophe Insurance Pool (TCIP) English Annual Reports (2012), available at http://www.tcip.gov.tr/content/annualReport/2012_Annual_Report_DASK.pdf.

[101]   Youbaraj Paudel, *A Comparative Study of Public–Private Catastrophe Insurance Systems: Lessons from Current Practices*, 37 The Geneva Papers on Risk and Insurance-Issues and Practice 257 (2012).

[102]   Paul A. Raschky and Hannelore Weck-Hannemann, *Charity Hazard—A Real Hazard to Natural Disaster Insurance?*, 7 Environmental Hazards 321 (2007).

[103]   Anastasia Telesetsky, *Insurance as a Mitigation Mechanism: Managing International Greenhouse Gas Emissions through Nationwide Mandatory Climate Change Catastrophe Insurance*, 27 Pace Environmental Law Review 691 (2010). See also Omri Ben-Shahar and Kyle D. Logue, *Outsourcing Regulation: How Insurance Reduces Moraz Hazard*, 111 Michigan Law Review 197 (2012) ("Insurance arrangements—by using such tools as deductibles, exclusions, and experience rating—give private parties the incentive to reduce risks").

[104]   Howard Kunreuther, *Mitigating Disaster Losses through Insurance*, 12 Journal of Risk and Uncertainty 171 (1996).

[105]   Howard Kunreuther, *Mitigating Disaster Losses through Insurance*, 12 Journal of Risk and Uncertainty 171 (1996).

[106]   Youbaraj Paudel, *A Comparative Study of Public–Private Catastrophe Insurance Systems: Lessons from Current Practices*, 37 The Geneva Papers on Risk and Insurance-Issues and Practice 257 (2012).

diminishing their drawbacks. For example, multiyear insurance does not address consumers' reliance on government bailouts. According to the empirical study of catastrophe insurance in five Chinese provinces mentioned above, because of biases and misunderstanding, few residents will voluntarily purchase catastrophe insurance.[107] Mandatory insurance removes adverse selection by requiring everyone to purchase catastrophe insurance and thus decreases reliance on government compensation dramatically.

As mentioned above, China suffers some of the most severe national catastrophes in the world. If mandatory multiyear insurance existed, it would reduce the financial burden on governments to compensate disaster victims. If premiums of mandatory multiyear insurance sufficiently reflect risks, the insurance industry can play an important role in covering catastrophe risks.

V.   CONCLUSION

Climate change, its implications for catastrophes, and the question of how losses caused by disasters can be covered are pressing topics that will attract more and more attention in coming years. Accordingly, how to sustainably distribute catastrophe risks is of pressing concern, especially for those countries and regions that have many natural disasters, such as the United States, China, and the European Union.

Insurance is one of mankind's greatest inventions, an extraordinarily useful tool to reduce risk.[108] Insurance is a good device to transfer risk for individuals. Policyholders can substitute a small certain cost, the premium, for a large uncertain financial loss. Compared with government intervention, insurance is better equipped to deal with catastrophe risks due to lower transaction costs, lower adverse selection, and more efficiency as a result of competitive markets. Furthermore, when facing catastrophe risks, mandatory multiyear insurance, which combines the

---

[107]   L. Tian and Y. Zhang, *Influence Factors of Catastrophe Insurance Demand in China—Panel Analysis in a Case of Insurance Premium Income of Five Provinces [Woguo Juzai Baoxian Xuqiu Yingxiang Yinsu Shizheng Yanjiu: Jiyu Wusheng Bufen Baofei Shouru Mianban Yanjiu]*, 26 Wuhan University of Technology (Social Science Edition) [Wuhan Ligong Daxue Xuebao (Shehui Kexue Ban)] 175 (2013).

[108]   Howard Kunreuther, Mark V. Pauly and Stacey McMorrow, Insurance and Behavioral Economics: Improving Decisions in the Most Misunderstood Industry 13 (2013).

merits of mandatory insurance and multiyear insurance while diminishing their drawbacks, deserves more attention.

There is no doubt that insurance could and should play an important role in covering catastrophic disasters, especially in China, where private insurance still plays a limited role in distributing and mitigating catastrophe risks.[109] Taking into account that China suffers some of the most severe national catastrophes in the world, mandatory multiyear insurance in which premiums sufficiently reflect risks may be worth considering.

---

[109]   The catastrophe disaster loss covered by insurance is less than 5 percent in China. This is much less than the international normal loss ratio of 36 percent. See *CIRC is Promoting to Stimulate Regulation on Catastrophe Insurance*, Economic Observer (October 2013), available at http://www.iic.org.cn/D_newsDT/newsDT_read.php?id=106741.

3. Mitigation of climate-change risks and regulation by insurance

I. INTRODUCTION

Climate change is one of the most fundamental challenges of our time.[1] The controversy about climate change cuts across scientific theory to litigation. Most scientists who contributed to the United Nations Intergovernmental Panel on Climate Change (IPCC) 2014 report believe that global climate change is occurring on a significant scale.[2] "[Climate change] leads to changes in the frequency, intensity, spatial extent, duration, and timing of extreme weather and climate events, and can result in unprecedented extreme weather and climate events."[3] It seems very likely that damages resulting from climate change might mount as high as

---

[1] The terms "climate change" and "global warming" are used interchangeably throughout this book. Climate change is considered the phenomenon by which human activity has altered the Earth's atmosphere. See 2 Handling the Land Use Case § 42:1 (3rd edn).

[2] Intergovernmental Panel on Climate Change, United Nations Environmental Program, *Climate Change 2014: Impacts, Adaptations, and Vulnerability* 1133, 1136 (2014), available at http://www.ipcc.ch/report/ar5/wg2/. This report states that the effects of climate change are already occurring on all continents and across the oceans. It has been produced by 309 coordinating lead authors and review editors, drawn from 70 countries. It also enlists the help of 436 contributing authors, and 1,729 expert and government reviewers. In addition, an extensive dataset of 1,372 climate researchers and their publication and citation data shows that 97 to 98 percent of the climate researchers most actively publishing in the field support the tenets of anthropogenic climate change outlined by the Intergovernmental Panel on Climate Change. See William R.L. Anderegga et al., *Expert Credibility in Climate Change*, PNAS: 12107–12109, Vol. 107, No.27 (2010).

[3] IPCC, *Summary for Policymakers* in Managing the Risks of Extreme Events and Disasters to Advance Climate Change Adaptation 3–21 (C.B. Field, V. Barros, T.F. Stocker, D. Qin, D.J. Dokken, K.L. Ebi, M.D. Mastrandrea, K.J. Mach, G.-K. Plattner, S.K. Allen, M. Tignor and P.M. Midgley eds), A Special Report of Working Groups I and II of the Intergovernmental Panel on Climate Change (Cambridge University Press, Cambridge, UK and New York, NY, USA, 2012).

one trillion dollars annually by 2040.[4] In *Massachusetts v Environmental Protection Agency (EPA)*,[5] the Supreme Court of the United States for the first time recognized that greenhouse gases (GHG) could have "a significant, disruptive impact on our climate".[6]

The extraordinary growth of GHG in China represents the single greatest challenge to global climate change efforts in coming decades. Without a significant contribution from China, efforts to find a solution to global climate change are unlikely to succeed.[7] Meanwhile, China suffers from the adverse consequences of climate change, and the impact has closely followed the global trend. Floods, heavy rainfall, landslides and many other climate hazards are likely to increase dramatically.[8] The increasing frequency and intensity of catastrophe disasters will no doubt aggravate the vulnerability of the socio-economic development of China.[9]

It has been recognized that two factors may increase climate-change risks: (a) the rising GHG emissions that will increase the frequency and intensity of climate hazards,[10] and (b) the accumulation of value-at-risk,

---

[4]   United Nations Environment Programme Finance Initiative Climate Change Working Group, *Adaptation and Vulnerability to Climate Change: The Role of the Finance Sector 14* (2006), available at http://www.unepfi.org/fileadmin/documents/ CEO_briefing_adaptation_vulnerability_2006.pdf ("It seems very likely that the [sic] there will be a 'peak' year that will record costs over 1 trillion USD before 2040").

[5]   549 U.S. 497 (2007).

[6]   Joseph MacDougald and Peter Kochenburger, *Insurance and Climate Change*, 47 John Marshall Law Review 719, 720–721 (2013).

[7]   Alex Wang, *Climate Change Policy and Law in China* in Oxford Handbook of International Climate Change Law 635–669 (Cinnamon P. Carlarne, Kevin R. Gray, and Richard Tarasofsky eds., 2016).

[8]   Based on the regression analysis of natural disaster occurrence and average global temperature from 1980 to 2010, the frequency of epidemics, extreme temperature, floods and storms was estimated to increase by 506 times per year if the average global temperature increases by 1°C. See XB Pan et al., *Natural Disaster Occurrence and Average Global Temperature*, 4 Disaster Adv 61 (2011).

[9]   Sha Chen, Zhongkui Luo and Xubin Pan, *Natural Disasters in China: 1900–2011*, 69 Natural Hazards 1597 (2013).

[10]   "Scientific consensus has established with significant confidence a link between CO2 emissions and human-induced global warming". See Daniel J. Grimm, *Global Warming and Market Share Liability: A Proposed Model for Allocating Tort Damages among CO2 Producers*, 32 Columbia Journal of Environmental Law 209, 212 (2007). A similar statement has been confirmed by US Supreme Court decision in *Massachusetts v Environmental Protection Agency*, 549 U.S. 497 (2007). Commentators Webster et al. and Knutson and Tuleya highlight that "a doubling of CO2 may increase the frequency of the most intense cyclones although attribution of the 30-year trend to global warming would require a longer global data record". See P.J. Webster, G.J. Holland, J.A. Curry

such as the concentration of the world's population and property, in vulnerable areas.[11] Therefore, mitigation of climate-change risk involves not only human intervention to reduce the GHG emissions but also prevention of potential losses caused by climate hazards. Among the many solutions to risk mitigation, insurance has received increased attention due to its emphasis on risk management and its regulatory function in influencing policyholders' behaviors. In other words, insurance could not only compensate the victims of climate hazards, which is also referred to as adaptation[12] to climate change, but also reduce climate-change risks.

This chapter examines the ability of two types of insurance—liability insurance and catastrophe insurance—to regulate and thus help mitigate climate-change risks, including risks from rising GHG emissions, and considers the lessons for China. First, I examine the connection between climate change and insurance. Next, I focus specifically on the insurability of climate-related liability risk and catastrophe risk, respectively. Then, I explore why regulation by liability insurance, whose appropriateness relies on the efficiency of tort-based climate-change litigation for loss mitigation, is infeasible, especially in China. Finally, I compare the liability insurance model with the catastrophe insurance model and then propose a catastrophe insurance-based private-public model for China.

II.   CLIMATE CHANGE AND INSURANCE

A.   The Impact of Climate Change on the Insurance Industry

Inevitably, climate change does and will continue to affect the insurance industry, whose function is to shield individuals and businesses from

---

and H.R. Chang, *Changes in Tropical Cyclone Number, Duration, and Intensity in a Warming Environment*, 309 Science 1844 (2005); Thomas R. Knutson and R.E. Tuleya, *Impact of CO2-induced Warming on Simulated Hurricane Intensity and Precipitation: Sensitivity to the Choice of Climate Model and Convective Parameterization*, 17 Journal of Climate 3477 (2004).

[11]   Value-at-risk refers to the increase in the value, such as asset values, exposed to natural hazards. See Arthur Charpentier, *Insurability of Climate Risks*, 33 The Geneva Papers on Risk and Insurance—Issues and Practice 91 (2008); Howard Kunreuther and Erwann Michel-Kerjan, *Climate Change, Insurability of Large-Scale Disasters, and the Emerging Liability Challenge*, 155 University of Pennsylvania Law Review 1795 (2007).

[12]   Adaptation generally involves recognizing that there is climate change and then changing polices to "make the best of it" and do something different to adapt to it.

risk.[13] The American International Group (AIG), Swiss Re, Lloyd's of London, and other leading insurers and reinsurers all identify climate change as a major threat to global risk management.[14] Major types of insurance products have been affected. First, the most widely recognized impact is catastrophic property losses caused by extreme weather hazards.[15] The level of economic losses, including buildings, houses, factories, and business interruption, increases dramatically due to both climate change and an increasing concentration of the world's population in vulnerable areas.[16] For example, in 1992 Hurricane Andrew caused havoc for catastrophe insurers.[17] The total paid claims were $15.5 billion for that event, and more than 10 insurers went into insolvency.[18] Some property insurers that suffered from extensive catastrophe loss claims even filed lawsuits against the state for its negligence in addressing climate change.[19] Similarly, impacts of climate change on the liability insurance sector are beginning to appear.[20] Liability claims related to climate change are

---

[13]   James M. Davis, *Global Warming Litigation—Implications for Insurance Coverage*, New Appleman on Insurance: Current Critical Issues in Insurance Law 1 (2007).

[14]   AIG, AIG's Policy and Programs on Environment and Climate Change 1, 1 (2009), available at http://media.corporate-ir.net/media_files/irol/76/76115/aig_climate_change_updated.pdf; Trevor Maynard, *Climate Change: Impacts on Insurers and How They Can Help with Adaptation and Mitigation*, 33 The Geneva Papers on Risk and Insurance—Issues and Practice 140 (2008).

[15]   Christina Ross, Evan Mills and Sean B. Hecht, *Limiting Liability in the Greenhouse: Insurance Risk-Management Strategies in the Context of Global Climate Change*, 26A Stanford Environmental Law Journal 251, 252 (2007).

[16]   Howard C. Kunreuther and Erwann Michel-Kerjan, *Climate Change, Insurability of Large-Scale Disasters, and the Emerging Liability Challenge*, 155 University of Pennsylvania Law Review 1795, 1806 (2007).

[17]   Rawle O. King, *Hurricanes and Disaster Risk Financing Through Insurance: Challenges and Policy Options*, Congressional Research Service, R132825 (2005), available at http://www.policyarchive.org/handle/10207/2378.

[18]   Lynne McChristian, *Hurricane Andrew and Insurance: The Enduring Impact of an Historic Storm*, 2 (2012), available at http://www.iii.org/sites/default/files/paper_HurricaneAndrew_final.pdf.

[19]   For example, in May 2014, subsidiaries of Farmers Insurance filed class-action lawsuits against a number of communities in the Chicago area because those communities had not done enough to prepare for the previous year's heavy rains and widespread flooding, which could have been anticipated due to global warming. They argued that extensive property damage was caused by these extreme weather hazards. See Climate Change: Insurance Issues 2014, available at http://www.iii.org/issue-update/climate-change-insurance-issues.

[20]   The relevant categories of liability insurance surrounding climate-change impact include commercial general liability, product liability, environmental liability, professional liability (directors' and officers' liability insurance), political risk

already being filed.[21] Hundreds of climate-change cases have been raised in the courts of 18 countries on six continents, and these are projected to impose material costs on liability insurers.[22] Additionally, climate change presents public health concerns relevant to life-health insurance lines. For example, extreme weather may cause possible outbreaks of respiratory and infectious diseases that increase the severity of personal risks including illness, disability, and death.[23]

Among different lines of insurance products, non-life insurers have to make much more effort than life insurers to address the impacts of climate change.[24] Property losses and liability litigation, especially damage to structures caused by extreme weather hazards, receive the largest attention.[25] Meanwhile, much less consideration has been given to life exposures.[26] Therefore, this chapter will mainly discuss mitigation of climate-change risks through liability insurance and catastrophe insurance.

---

liability concerning new government policies, and personal and commercial vehicle liability. See Christina Ross, Evan Mills and Sean B. Hecht, *Limiting Liability in the Greenhouse: Insurance Risk-Management Strategies in the Context of Global Climate Change*, 26A Stanford Environmental Law Journal 251 (2007).

[21]   Christina Ross, Evan Mills and Sean B. Hecht, *Limiting Liability in the Greenhouse: Insurance Risk-Management Strategies in the Context of Global Climate Change*, 26A Stanford Environmental Law Journal 251, 258 (2007).

[22]   Jacqueline Peel and Hari M. Osofsky, Climate Change Litigation 1–2 (2015).

[23]   Sean B. Hecht, *Climate Change and the Transformation of Risk: Insurance Matters*, 55 UCLA Law Review 1559, 1575–1576 (2008).

[24]   Non-life insurers develop modeling techniques that enable them to predict the increasingly visible signs of climate change risk. For example, by using state-of-the-art modeling techniques, several important findings have been released on the financial impact on insured risk of inland floods in Great Britain, winter windstorms in the UK, and typhoons in China. However, life insurers may not be prepared for the dramatic changes in mortality and morbidity relating to illness and death caused by extreme weather. See Hsin-Chun Wang, *Adaptation to Climate Change and Insurance Mechanism: A Feasible Proposal Based on a Catastrophe Insurance Model for Taiwan*, 9 NTU Law Review 317, 322 (2014).

[25]   For example, US-based insurers' knowledge of climate-change impacts has been largely focused on property and casualty (P&C) insurance lines, especially on damage to fixed structures. See Christina Ross, Evan Mills and Sean B. Hecht, *Limiting Liability in the Greenhouse: Insurance Risk-Management Strategies in the Context of Global Climate Change*, 26A Stanford Environmental Law Journal 251, 258 (2007). In addition, most climate change litigations involve involve non-life insurers rather than life insurers. See Jacqueline Peel and Hari M. Osofsky, Climate Change Litigation 1–4 (2015).

[26]   Christina Ross, Evan Mills and Sean B. Hecht, *Limiting Liability in the Greenhouse: Insurance Risk-Management Strategies in the Context of Global Climate Change*, 26A Stanford Environmental Law Journal 251, 258 (2007). Life

B.   Why Insurance Could Theoretically Mitigate Climate-Change Risks

As a well-known professional risk management mechanism, insurance plays an important role in mitigating both property and liability risks.[27] For example, the 2012 report of the IPCC identified insurance as a risk mitigation tool for extreme weather events.[28] Because they assume both liability and property risks, in theory insurers have the incentives and the capacity to mitigate risks and the resulting losses.[29] Scholars of regulation-by-insurance (also called insurance-as-governance) propose several theoretical explanations for the function of insurance in risks mitigation.

Some scholars assert that, in the modern state, insurers often perform behavior-control functions and create incentives for policyholders to mitigate risks.[30] For example, through insurance rate classification, liability insurers can charge experience-rated premiums and thus induce policyholders to act more carefully than they would otherwise.[31] In practice, insurance laws' reluctance to prohibit rate classification based on controllable characteristics supports insurers' behavior-control functions.

---

exposures generally refer to personal risks like illness, disability and death. See Hsin-Chun Wang, *Adaptation to Climate Change and Insurance Mechanism: A Feasible Proposal Based on a Catastrophe Insurance Model for Taiwan*, 9 NTU Law Review 317, 322 (2014).

[27]   Howard C. Kunreuther, *Linking Insurance and Mitigation to Manage Natural Disaster Risk* in Handbook of Insurance 593 (Georges Donnie ed., 2000).

[28]   IPCC, *Managing the Risks of Extreme Events and Disasters to Advance Climate Change Adaptation* in A Special Report of Working Groups I and II of the Intergovernmental Panel on Climate Change (C.B. Field et al., Cambridge, UK and New York, NY, USA: Cambridge University Press, 2012).

[29]   "If an insurer can lower its premiums by lowering its risk of paying claims, it can undercut its competitors by charging lower premiums, thereby attracting more business. Marketplace considerations, rather than altruism, drive insurers to reduce risk". See John Aloysius Cogan Jr, *The Uneasy Case for Food Safety Liability Insurance*, 81 Brooklyn Law Review 1495 (2016). With respect to capacity, insurers have many techniques to reduce risk, and the consensus is that these techniques work reasonably well. See Tom Baker and Peter Siegelman, *The Law & Economics of Liability Insurance: A Theoretical and Empirical Review* in Research Handbook on the Economics of Tort 169 (Jennifer Arlen ed., 2013).

[30]   Kenneth S. Abraham, *Four Conceptions of Insurance*, 161 University of Pennsylvania Law Review 653, 685 (2013); Tom Baker and Jonathan Simon, *Embracing Risk* in Embracing Risk: The Changing Culture of Insurance and Responsibility 1, 13 (Tom Baker and Jonathan Simon eds., 2002); Jeffrey W. Stempel, *The Insurance Policy as Social Instrument and Social Institution*, 51 William & Mary Law Review 1489, 1498–1501 (2010).

[31]   For example, *Can You Afford Not to Be in Good Hands?*, Allstate, available at https://www.allstate.com/auto-insurance/auto-insurance-comparison.aspx.

Some scholars claim that, compared to the state, insurers have the capacity to manage moral hazard because of both superior information and competition.[32] Besides ex post indemnification, insurance uses regulatory techniques such as risk-based premiums, deductibles, exclusions, and loss-reduction services, to give policyholders the incentive to reduce risks and invest in prevention measures.[33]

In theoretical terms, risk-based pricing is regarded as the central approach of insurers to risk mitigation.[34] Insurers' premium-setting processes give policyholders the financial incentive to conduct mitigation measures. In theory, the insurance premium is based on the expected overall losses, derived by multiplying loss probability by loss severity.[35] Reducing either the probability or the severity of loss may lower the premium. As long as such reduction cost is lower than the discount of the premium, policyholders are likely to undertake mitigation.[36] However, if the loss probability and loss severity are too high, insurers may refuse to underwrite in the first place.[37]

What has been explained previously about insurance in general also

---

[32] Moral hazard is the tendency to exercise less care to avoid losses. See Tom Baker, *On the Genealogy of Moral Hazard*, 75 Texas Law Review 237 (1996). By utilizing the methodologies of actuarialism, private contracting, and ex post claim investigation, insurers can easily collect customers' purchasing information, thereby replacing government. See Omri Ben-Shahar and Kyle D. Logue, *Outsourcing Regulation: How Insurance Reduces Moral Hazard*, 111 Michigan Law Review 197 (2012). It is generally believed that insurance markets tend to be highly competitive with respect to price. See Daniel Schwarcz, *Regulating Consumer Demand in Insurance Markets*, 3 Erasmus Law Review 23, 43 (2010).

[33] Omri Ben-Shahar and Kyle D. Logue, *Outsourcing Regulation: How Insurance Reduces Moral Hazard*, 111 Michigan Law Review 197 (2012); Tom Baker and Rick Swedloff, *Regulation by Liability Insurance: From Auto to Lawyers Professional Liability*, 60 UCLA Law Review 1412 (2013); Haitao Yin, Howard Kunreuther and Matthew White, *Risk-Based Pricing and Risk-Reducing Effort: Does the Private Insurance Market Reduce Environmental Accidents?*, 54 The Journal of Law and Economics 325 (2011); Steven Shavell, *On the Social Function and Regulation of Liability Insurance*, 25 The Geneva Papers on Risk and Insurance—Issues and Practice 166 (2000).

[34] This is risk mitigation, plain and simple. See Tom Baker and Rick Swedloff, *Regulation by Liability Insurance: From Auto to Lawyers Professional Liability*, 60 UCLA Law Review 1412, 1419 (2013).

[35] Peter Molk, *Private Versus Public Insurance for Natural Hazards: Individual Behavior's Role in Loss Mitigation* in Risk Analysis of Natural Hazards 265 (Paolo Gardoni et al. eds., 2015).

[36] Peter Molk, *Private Versus Public Insurance for Natural Hazards: Individual Behavior's Role in Loss Mitigation* in Risk Analysis of Natural Hazards 265 (Paolo Gardoni et al. eds., 2015).

[37] This also serves as a gatekeeping function. See Tom Baker and Thomas O. Farrish, *Liability Insurance and the Regulation of Firearms* in Suing the

applies to liability and catastrophe insurance. In the case of liability insurance, which is designed to defend and indemnify insured emitters of GHGs in climate-change-related tort lawsuits, insurers could charge a lower premium for the insured emitter who reduces GHG emissions, or refuse to underwrite that risk if the insured is emitting too much. In the case of catastrophe insurance, which is designed to compensate victims of extreme weather hazards, insurers could, through higher premiums, deter the insured from locating in higher risk areas as compared to lower risk areas.[38] Therefore, theoretically, insurance could help reduce GHG emissions, reduce value-at-risk, and thus realize the mitigation of climate-change risks. While promising in theory, this is not always true in practice. Whether insurance could send these valuable price signals for climate-change risk mitigation depends on a number of practical issues. In the following discussion, I will examine how liability insurance and catastrophe insurance function as risk mitigation in practice.

III. INSURABILITY OF CLIMATE-CHANGE RISKS

Before answering the question of how insurance can function as risk mitigation, first we need to revisit the issue whether climate-change risks are insurable, which concerns the willingness of insurance companies to insure climate-related liability and catastrophe risk.[39] *Climate-change risks* is a generic term applied to a range of risks related to climate change. This chapter focuses on liability risk and property risk, and the following discussion will address the insurability of climate-change risks in the line of liability insurance (liability risk) and catastrophe insurance (property risk), respectively.

A.   Insurability of Climate-change Liability Risk

People worry that climate-change claims might be the "next asbestos" for insurers. Asbestos claims were some of the first mass tort cases to push

---

Gun Industry: A Battle at the Crossroads of Gun Control and Mass Torts 294 (Timothy D. Lytton ed., 2008).

[38]   Swenja Surminski, *The Role of Insurance in Reducing Direct Risk: The Case of Flood Insurance*, 7 International Review of Environmental and Resource Economics 241, 264 (2013).

[39]   Hsin-Chun Wang, *Adaptation to Climate Change and Insurance Mechanism: A Feasible Proposal Based on a Catastrophe Insurance Model for Taiwan*, 9 NTU Law Review 317, 323 (2014).

the boundaries of liability insurance insurability and coverage.[40] Many describe climate-change liability risk as an "emerging risk". When insurers initially underwrote climate-change risks, they were perhaps not aware of all the potential impacts. But now those risks are perceived as potentially significant, and even possibly the greatest risk to the property/casualty insurance industry.[41] In defining whether climate-change liability risks are insurable, the insurance literature identifies certain basic requirements to be considered:

(1) Actuarial estimation requires that the insurers can identify, quantify, and estimate the frequency and severity of risks and the resulting losses.
(2) A causal relationship requires that the causes of losses must be directly assignable and allocable to the insured as the subject of liability.
(3) Randomness requires that the materialization of the risk must be random, unintended, and unexpected.[42]

While these requirements present challenges for liability insurers, arguments that climate-change liability risk is uninsurable may not always be convincing.

First, liability insurers may worry that climate change presents uncertainty as to the intensity and frequency of natural disasters[43] and that increasing litigation could cause substantial financial losses to insurers

---

[40] Tom Baker and Kyle D. Logue, Insurance Law and Policy: Cases and Materials 330 (2013).

[41] Christina M. Carroll et al., Climate Change and Insurance 21 (2012); *Climate Change No. 1 in Top 10 Risks Facing the Insurance Industry*, Insurance Journal (March 12, 2008), available at http://www.insurancejournal.com/news/national/2008/03/12/88138.htm.

[42] Baruch Berliner, *Large Risks and Limits of Insurability*, 10 The Geneva Papers on Risk and Insurance—Issues and Practice 313 (1985); J. Spihler, Report of Emerging Risks: A Challenge for Liability Underwriters 33 (2009). See also Arthur Charpentier, *Insurability of Climate Risks*, 33 The Geneva Papers on Risk and Insurance—Issues and Practice 91 (2008); Hsin-Chun Wang, *Adaptation to Climate Change and Insurance Mechanism: A Feasible Proposal Based on a Catastrophe Insurance Model for Taiwan*, 9 NTU Law Review 317 (2014).

[43] The uncertainty of climate change risk will significantly affect the insurability of liability relating to tort litigation. It is inevitable that liability insurers will have to pay enormous litigation costs on behalf of the insured. See Hsin-Chun Wang, *Adaptation to Climate Change and Insurance Mechanism: A Feasible Proposal Based on a Catastrophe Insurance Model for Taiwan*, 9 NTU Law Review 317 (2014).

because insurers' duty to defend[44] is much broader than their duty to indemnify, at least in the United States. However, neither the size of the risk nor potential loss estimates has prevented successful insurance operations in the past.[45] Looking back at insurance history, we can see plenty of examples of insurance against catastrophe losses that the insurers could not predict in advance.[46] One example is commercial satellite insurance and commercial aircraft insurance. These insurance products both involve huge losses, and at the initial stage of underwriting, insurers had no historical data to assess the intensity and the frequency of losses.[47] Even considering that these risks are still smaller than those from climate change, insurance history is full of what people in the insurance trade call *assessment insurance*.[48] Assessment insurance means that insurers no longer face the same budget constraints that they did in the past.[49]

Second, there must be an actual causal relationship between the GHGs emitted by the insured and the resulting property damages or physical injuries. If not, the liability insurer has a strong argument for denying coverage. Causal uncertainty between climate change and GHG emitters will affect the efficiency of tort-based climate-change litigation.[50] It is

---

[44]   In a liability insurance policy, "the duty to defend" clause means that the insurers agree not only to pay tort judgments against the policyholder but also to defend any lawsuits brought by plaintiffs against the policyholder even if the suit is groundless, false or fraudulent. In this sense, liability insurance in effect becomes "litigation insurance". See Jeffrey W. Stempel, *Insurance and Climate Change Litigation*, in Adjudicating Climate Change: Sub-National, National and Supra-National Approaches 235 (William C.G. Burns and Hari M. Osofsky eds., 2009).

[45]   Dwight Jaffee and Thomas Russell, *Catastrophe Insurance, Capital Markets, and Uninsurable Risks*, 64 Journal of Risk and Insurance 205, 207 (1997).

[46]   Tom Baker, *Embracing Risk, Sharing Responsibility*, 56 Drake Law Review 561, 569 (2008).

[47]   Karl Borch, Economics of Insurance 315 (1990); Dwight Jaffee and Thomas Russell, *Catastrophe Insurance, Capital Markets, and Uninsurable Risks*, 64 Journal of Risk and Insurance 205, 207 (1997).

[48]   "With assessment insurance, the insurer has the ability to come back and collect more after a loss to help people who need it if the insurance fund runs dry". See Tom Baker, *Embracing Risk, Sharing Responsibility*, 56 Drake Law Review 561, 569 (2008).

[49]   For example, with Florida's Citizens Property Insurance Corporation, citizens can secure emergency funding for catastrophic losses that exceed the corporation's own reserves under the assessment process. Citizens can impose a tax on all citizens' policyholders. Part of this assessment/tax is collected up front, and part is spread out over a number of years, until the deficit is paid. See Omri Ben-Shahar and Kyle Logue, *The Perverse Effects of Subsidized Weather Insurance*, 68 Stanford Law Review 571 (2016).

[50]   This will be discussed in more detail in section III.A.1 below.

still a matter of debate whether large GHG emitters—such as fossil fuel companies, power plants, and automobile manufacturers—can be liable as potential tortfeasors for extreme weather events and the rise in sea levels.[51] Considering this causal uncertainty, we cannot consider climate-change liability risk entirely uninsurable because the probability of an insured loss is lower.

Third, the timing, magnitude, or location of extreme weather events cannot be known precisely in advance, and thus there may be both a demand for such insurance and a willingness by insurers to underwrite such risks. The randomness of climate hazards also means that insureds (i.e. homeowners) have no control over such liability risks, which speaks to the problems of moral hazard and adverse selection.[52] Moral hazard and adverse selection are generally caused by asymmetric information, which exists where one party has more information than the other.[53] However, insureds have no more information than insurers in cases of climate hazards.

Retroactive liability, which refers to the fact that in the interval between the original tort and the claim for damages the standard of care applied by the courts may change, may endanger the insurability of long-tail risks.[54] Indeed, in relation to environment liabilities some state courts may require insurers (ex post) to extend coverage or pay out based on retroactive liability.[55] However, liability insurers could argue in response that emitting

---

[51]   David A. Grossman, *Warming up to a Not-So-Radical Idea: Tort-based Climate Change Litigation*, 28 Columbia Journal of Environmental Law 1, 22–33 (2003).

[52]   Dwight Jaffee and Thomas Russell, *Catastrophe Insurance, Capital Markets, and Uninsurable Risks*, 64 Journal of Risk and Insurance 205, 206 (1997). Adverse selection means that those with the highest risk are most likely to buy the insurance. See Véronique Bruggeman, Compensating Catastrophe Victims: A Comparative Law and Economics Approach 64–66 (2010).

[53]   Véronique Bruggeman, Compensating Catastrophe Victims: A Comparative Law and Economics Approach 64–66 (2010).

[54]   Michael Faure and Paul Fenn, *Retroactive Liability and the Insurability of Long-Tail Risks*, 19 International Review of Law and Economics 487, 490–493 (1999). Simply speaking, long-tail risk refers to the risks arising out of the latent nature of certain insured events.

[55]   In some cases, legislators explicitly opted for a retroactive liability regime in the area of environmental liability; judges held that "a particular behavior (like dumping toxic waste) was already considered wrongful at the moment when the act happened (e.g., twenty-five years ago) even though it may be doubtful that this actually was the case". See Michael G. Faure, *Insurability of Damage Caused by Climate Change: A Commentary*, 155 University of Pennsylvania Law Review 1875, 1877 (2007).

GHGs is not like dumping toxic waste, which argument has been affirmed in findings of liability in some cases.[56] If emitting GHGs was not wrongful at the time of the act of emission it may not justify an action in tort, and thus the courts may not apply such retroactive case law.[57]

B.   Insurability of Climate-change Catastrophe Risk

Catastrophe insurance is a type of first-party insurance that is designed to indemnify the insured for direct losses resulting from a covered peril, the prospective victims being paid by the insurers once a covered event occurs.[58] Climate-change risks, such as extreme weather or rises in sea levels present significant challenges to catastrophe insurance due to the huge losses that result.[59] Catastrophes are often called "uninsurable risk", and insurers are likely to flee from underwriting.[60] The insurance literature often identifies three factors defining an uninsurable catastrophe risk:

(1)   Ambiguity of risk, also called uncertainty: the inability to identify and quantify probabilities of predicted losses with sufficient precision.
(2)   Losses and insolvency: concern that the largest possible loss could threaten insurers' solvency.
(3)   Appetite: insurers lack the desire to underwrite climate-change risks at a price policyholders are willing to pay.[61]

---

[56]   For examples and a discussion of the retroactive case law, see Gerrit Betlem and Michael Faure, *Environmental Toxic Torts in Europe: Some Trends in Recovery of Soil Clean-Up Costs and Damages for Personal Injury in the Netherlands, Belgium, England and Germany*, 10 Georgetown International Environmental Law Review 855 (1998).

[57]   Michael G. Faure, *Insurability of Damage Caused by Climate Change: A Commentary*, 155 University of Pennsylvania Law Review 1875, 1880 (2007).

[58]   George J. Couch and Ronald A. Anderson, Couch Cyclopedia of Insurance Law § 1:61 (2nd edn, 1984); Robert Jerry and Douglas R. Richmond, Understanding Insurance Law 45–46 (2012).

[59]   For example, in 2005 Hurricane Katrina caused about $48 billion in catastrophe-related insured losses. After Hurricane Katrina, a lot of insurance firms cut back their coverage in coastal areas. Sean B. Hecht, *Climate Change and the Transformation of Risk: Insurance Matters*, 55 UCLA Law Review 1559, 1582 (2008).

[60]   Besides huge losses caused by climate hazards, catastrophic risks require insurers to hold large amounts of liquid capital, but institutional factors (such as accounting, tax, and takeover risk) make insurers reluctant to do this. See Dwight Jaffee and Thomas Russell, *Catastrophe Insurance, Capital Markets, and Uninsurable Risks*, 64 Journal of Risk and Insurance 205, 206 (1997).

[61]   Celine Herweijer, Nicola Ranger and Robert E.T. Ward, *Adaptation to Climate Change: Threats and Opportunities for the Insurance Industry*, 34 The

This chapter argues, however, that despite these concerns, climate-change catastrophe risks are, to some extent, insurable.

First, climate change leads to an increase in the uncertainty associated with the frequency and severity of extreme weather events. When there is "too much" uncertainty, the exposure becomes unmeasurable and unquantifiable, and thus uninsurable. Nonetheless, with the steadily growing body of data on catastrophe events and the development of catastrophe models that could help estimate potential damages of a catastrophe, natural catastrophe risk is evolving away from a highly uncertain line of business.[62]

Second, like liability insurance, discussed above, catastrophe insurance involves similar concerns and arguments about the potential magnitude of catastrophe losses. To address the concern of insolvency, outside capital could supplement catastrophe insurers' capacity. Reinsurance and insurance-linked securitization (i.e., catastrophe bonds)[63] could provide additional sources to primary insurers. In addition and as a last resort, the government could also contribute to solving this problem.[64]

Third, the asymmetric information problem that affects insurers' underwriting may not be as severe as presumed. The so-called moral hazard might be minimal because the risk of catastrophic loss is not the private information of the insured.[65] In addition, many insurers consult with

---

Geneva Papers on Risk and Insurance—Issues and Practice 360 (2009); Sean B. Hecht, *Climate Change and the Transformation of Risk: Insurance Matters*, 55 UCLA Law Review 1559, 1565, 1580–1581 (2008); Arthur Charpentier, *Insurability of Climate Risks*, 33 The Geneva Papers on Risk and Insurance—Issues and Practice 91 (2008); Howard C. Kunreuther and Erwann O. Michel-Kerjan, *Climate Change, Insurability of Large-Scale Disasters, and the Emerging Liability Challenge*, 155 University of Pennsylvania Law Review 1795, 1813 (2007); Baruch Berliner, Limits of Insurability of Risks 42–43 (1982).

[62] Swiss Re, Innovating to Insure the Uninsurable 14 (2005).

[63] Reinsurance can be understood simply as insurer's insurance. Insurers have protected themselves through private reinsurance contracts whereby portions of their losses from a catastrophic disaster are covered by some type of reinsurance arrangement. Insurance-linked securitization could be regarded as the process of transferring insurance risks from insurers and conveying them to third parties through tradable securities. Catastrophe bonds also called "cat bonds" or "Act of God bonds", and are the most prominent and popular form of insurance-linked securities.

[64] The government has a deep credit capacity due to its ability to raise money through tax or borrow money by issuing debt far more readily than private insurers or reinsurers. See Louis Kaplow, *Incentives and Government Relief for Risk*, 4 Journal of Risk and Uncertainty 167 (1991).

[65] Dwight Jaffee and Thomas Russell, *Catastrophe Insurance, Capital Markets, and Uninsurable Risks*, 64 Journal of Risk and Insurance 205, 206 (1997).

scientists on predicting climate-change risks and thus have the advantage of information.

IV.   REGULATION BY LIABILITY INSURANCE FOR MITIGATION OF GHG EMISSIONS?

While liability insurers in theory have the ability to mitigate GHG emissions through regulatory tools (and it may not be appropriate to declare climate-change liability risk to be uninsurable), it is unlikely that regulation by liability insurance will function effectively in practice. Through a comparative study of the United Sates and China, I will explore the possible reasons and identify specific obstacles hindering regulation by liability insurance in China.

A.   Tortious Liability and Liability Insurance Coverage

The rules of tort law and the resulting liability—if any—for climate-change risk play a major role in determining whether liability insurance could help mitigate GHG emissions. Any such liability would force GHG emitters to internalize the risk that insurers then regulate and manage.[66] If polluters face no liability for their behavior, there will be no demand for liability insurance in the first place. Thus, the current obstacles for the establishment of tortious liability need to be clearly understood.

1.   Obstacles to tortious liability

Causation uncertainty is an important issue in climate-change tortious cases, as claimants must show that their damages are caused by the defendants' actions.[67] In climate-change cases, as in many toxic tort cases, a legally sufficient causal relationship does not exist. Although the US Supreme Court has recognized a general causal link between climate

---

[66]   Internalizing the risk is contrary to insurance externality, which refers to the situation that "if insurers do not classify insureds into sufficiently narrow risk pools, insureds will in large measure externalize accident costs to their insurers". See Jon D. Hanson and Kyle D. Logue, *The First-party Insurance Externality: An Economic Justification for Enterprise Liability*, 76 Cornell Law Review 129, 131 (1990).

[67]   David A. Grossman, *Tort-Based Climate Litigation* in Adjudicating Climate Change: Sub-National, National and Supra-National Approaches 215 (William C.G. Burns and Hari M. Osofsky eds., 2009).

change and man-made GHGs emissions,[68] proof of proximate causation by individual emitters is generally not available.[69] No US court has found proximate cause between the emissions of an individual emitter and adverse effects (now and in the future) on the climate.

Causal uncertainty leads to other legal disputes that include the identification of tortfeasors and the determination of losses. Since everyone may emit some GHGs and contribute in some degree to climate change, climate-change victims should determine who is a viable defendant.[70] On the other hand, how to assess the damages due to climate change and determine the number of losses is extremely difficult. For example, GHG emissions that cause damages may have happened many years ago, and this factor may hinder loss assessment.[71]

In response to the causal uncertainty issue, some commentators propose using proportional liability theory and market share liability theory to determine whether the emitters should be liable for damages caused by climate change; however, neither provides an adequate solution to this problem.[72] Compensation based on proportional liability could be estimated on the increased risk of natural disasters caused by GHGs.[73] However, how to identify the culpable emitters and the appropriate amount of damages in response to the increased risk of climate-related disasters is still a practical difficulty.[74] Even it helps facilitate the plaintiffs in recovering proportional compensation, litigation costs involving proof

---

[68]   *Massachusetts v Environmental Protection Agency*, 549 U.S. 497, 504 (2007).

[69]   Hsin-Chun Wang, *Adaptation to Climate Change and Insurance Mechanism: A Feasible Proposal Based on a Catastrophe Insurance Model for Taiwan*, 9 NTU Law Review 317, 328 (2014).

[70]   Ina Ebert, *Legal Aspects of US Claims based on Greenhouse Gas Emissions* in Liability for Climate Change? Experts' Views on a Potential Emerging Risk 15 (Munich Re edited, 2010).

[71]   Ina Ebert, *Legal Aspects of US Claims based on Greenhouse Gas Emissions* in Liability for Climate Change? Experts' Views on a Potential Emerging Risk 15 (Munich Re edited, 2010).

[72]   David A. Grossman, *Warming up to a Not-So-Radical Idea: Tort-based Climate Change Litigation*, 28 Columbia Journal of Environmental Law 1, 28–31 (2003); Daniel J. Grimm, *Global Warming and Market Share Liability: A Proposed Model for Allocating Tort Damages Among CO2 Producers*, 32 Columbia Journal of Environmental Law 209, 211 (2007).

[73]   For example, if the frequency of one extreme weather event increased by 50 percent, the emitters would be held liable for 50 percent of the victims' losses. See Myles Allen, *Liability for Climate Change*, 421 Nature 891, 891–892 (2003).

[74]   Howard C. Kunreuther and Erwann Michel-Kerjan, *Climate Change, Insurability of Large-Scale Disasters, and the Emerging Liability Challenge*, 155 University of Pennsylvania Law Review 1795, 1835 (2007).

of the appropriate proportion would be huge.[75] In the case of market share liability, which was first referred in the case of *Sindell v Abbot Laboratories*[76] for fungible products, it "is premised on the notion that a defendant's contribution to the aggregate risk of harm should approximate the defendant's output".[77] By applying market share liability, the first big difference is that climate change is not fungible products. Furthermore, market share liability requires that "the entirety of the plaintiff's injury was caused by the defendants", while "the entirety of a plaintiff's injuries will not possibly have been caused only by those firms in the emissions market".[78] Therefore, how to define the accuracy and the scale of the emissions market on a global basis is not easy for victims.

Proof of causation is also required by several statutes in China. Article 2 of the Insurance Act requires a causal link between the risk and the loss.[79] A similar provision is found in Article 216(1) of the Maritime Code. Taken together, causation uncertainty and other obstacles suggest that, under current tort rules, tortious liability might not be established between GHG emissions of individual emitters and climate-change victims' losses.

2. Obstacles for liability insurance coverage

Even if a GHG emitter's tortious liability were established by a court, obstacles would remain to insurance coverage. The case *AES v Steadfast*—the first and only case directly addressing climate-change claims under a liability policy[80]—shows the difficulty of prevailing on a liability insurance claim for climate-change liability.

---

[75]   Hsin-Chun Wang, *Adaptation to Climate Change and Insurance Mechanism: A Feasible Proposal Based on a Catastrophe Insurance Model for Taiwan*, 9 NTU Law Review 317, 329 (2014).

[76]   607 P.2d 924, 936 (Cal. 1980).

[77]   Daniel J. Grimm, *Global Warming and Market Share Liability: A Proposed Model for Allocating Tort Damages Among CO2 Producers*, 32 Columbia Journal of Environmental Law 209, 215 (2007).

[78]   Daniel J. Grimm, *Global Warming and Market Share Liability: A Proposed Model for Allocating Tort Damages Among CO2 Producers*, 32 Columbia Journal of Environmental Law 209, 226–227 (2007).

[79]   Article 2 of the Insurance Act provides that "to indemnify the insured where property loss or damage *is caused as a result of* the occurrence of an insured event".

[80]   AES was one of the defendants in the case of *Native Village of Kivalina v ExxonMobil*. The plaintiff Kivalina sued numerous emitters of GHGs and attributed the impending destruction of its land to the defendants because of their emissions of large quantities of GHGs. See 696 F.3d 849 (9th Cir. 2012), cert denied 133 S.Ct. 2390 (2013). When sued, AES notified its liability insurer Steadfast Insurance Company of the duty to defend based on the provisions of Commercial General

A vital question for liability insurance coverage is whether the damage was caused by an occurrence. An "occurrence" is typically defined in US liability policies as "an accident, including continuous or repeated exposure to substantially the same general harmful conditions".[81] Do GHG emissions constitute an "occurrence"? The answer in the decision of *AES v Steadfast* is no. The court applied the "eight corners rule" and reasoned that "when the insured knows or should have known the consequences of his actions, there is no occurrence".[82] In other words, if the insured was aware of the GHG emissions and the resulting damage was foreseeable, there was objective intent, based on the perspective of a hypothetical "reasonable" person. The Virginia Supreme Court declined to adopt the insured's subjective intent standard in assessing "occurrence".[83]

Chinese law presents similar rules referred to as "indirect intention". Article 27 of the Insurance Act provides that "Where the assured or the insured intentionally causes the occurrence of an insured event, the insurer is entitled to terminate the contract and shall not be liable for indemnity payment of insurance benefits".[84] Furthermore, when the insured knows with substantial certainty that the harm would occur as a result of his action, his act meets the requirement of "indirect intention" in the legal theory of tort law. Indirect intention may entitle the insurer to reject the claim since the insured's act does "intentionally" cause damage.[85]

There are further issues beyond those raised in *AES v Steadfast*. For

---

Liability (CGL) policy. Steadfast stated in the Virginia Supreme Court it they did not have the duty to defend for three reasons: "(1) the Complaint did not allege 'property damage' caused by an 'occurrence'; (2) any alleged injury arose prior to the inception of Steadfast's coverage; (3) the claims fell within the scope of the pollution exclusion". The Virginia Supreme Court applied the "eight corners rule" and held that Steadfast did not have the duty to defend AES. See *AES v Steadfast Ins Corp*, 725 S.E.2d 532, 532 (Va. 2012).

[81]   D. Ehrich, D. Young Morrissey and D. Herbert, *Insurance Law for Climate Related Claims*, LexisNexis Global Climate Change Special Pamphlet Series (September 2010) § 1.06[2].

[82]   The so-called "eight corners rule" means the court determines whether the allegations in the underlying complaint come within the coverage provided by the policy by comparing the "four corners" of the underlying complaint with the "four corners" of the policy. See *AES v Steadfast Ins Corp*, 725 S.E.2d 532, 535 (Va. 2012).

[83]   Lee R. Russ, Couch on Insurance §103:27 (3rd edn, 2011) (noting dispute as to whether the "expected or intended" element of the term "occurrence" is interpreted objectively or subjectively).

[84]   The Insurance Act of PRC Article 27.

[85]   Tao Yu, *Liability Insurance* in Insurance Law in China 135, 166 (Johanna Hjalmarsson and Dingjing Huang eds., 2015).

example, the pollution exclusion is not addressed in that case. Whether carbon dioxide is a "pollutant" within Commercial General Liability (CGL) policy's pollution exclusion is hotly contested,[86] although insureds argue that carbon dioxide is an "omnipresent", "odorless and colorless gas" and thus should not be categorized as a pollutant.[87] This argument was challenged in the decision of *Massachusetts v EPA*. The US Supreme Court held that GHGs fell within the definition of "air pollutants" in section 302(g) of the Clean Air Act.[88] The Evironmental Protection Agency (EPA) adheres to this judgment that GHGs are "without doubt, physical chemical substances emitted into the ambient air".[89]

In the Chinese context, other legal doctrines impede the potential of liability insurance in risk mitigation. An insurer's duty to defend is a unique example. In the United States, the duty to defend, an essential part of a liability insurance policy, means that the insurers agree not only to pay tort judgments against the policyholder but also to defend the lawsuits brought by plaintiffs against the policyholder even if the suit is groundless, false, or fraudulent.[90] Under the duty to defend, liability insurers in the United States might be forced to become involved in climate-change litigation.[91] In China, however, insurance law only specifies the liability insurers' duty to indemnify but "imposes no duty to defend the insureds".[92] Although insurers have the right to participate in legal actions and issues of compensation, in practice few insurers do so.[93] Insurers are not bound by a duty to defend when claimants sue GHG emitters; therefore, they

---

[86] CGL policy normally includes a "pollution exclusion" clause. If carbon dioxide is a "pollutant", there will be no coverage for the resulting damages.

[87] Brief of Appellee at 40, *AES Corp v Steadfast Ins Co*, 715 S.E.2d 28 (Va. 2011) (No. 100764), 2010 WL 6893536.

[88] 549 U.S. 497 (2007).

[89] EPA, "Endangerment and Cause or Contribute Findings for Greenhouse Gases under the Section 202(a) of the Clean Air Act 80 (December 7, 2009) (quoting Steve Jones, *Virginia Supreme Court to Decide Insurance Coverage for Climate Change Suits*, Marten Law (June 2, 2011), http://www.martenlaw.com/newsletter/20110602-insurance-coverage-climate-change#_edn26.)

[90] Jeffrey W. Stempel, *Insurance and Climate Change Litigation* in Adjudicating Climate Change: Sub-National, National and Supra-National Approaches 235 (William C.G. Burns and Hari M. Osofsky eds., 2009).

[91] Jeffrey W. Stempel, *Insurance and Climate Change Litigation* in Adjudicating Climate Change: Sub-National, National and Supra-National Approaches 235 (William C.G. Burns and Hari M. Osofsky eds., 2009).

[92] Tao Yu, *Liability Insurance* in Insurance Law in China 135, 161 (Johanna Hjalmarsson and Dingjing Huang eds., 2015).

[93] Tao Yu, *Liability Insurance* in Insurance Law in China 135, 169 (Johanna Hjalmarsson and Dingjing Huang eds., 2015).

have few incentives to become involved in such cases, let alone regulate emitters' behaviors. Regulating emitters' behaviors ex ante is much less costly for insurers if the insureds produce more emissions and then are sued as a result.

B. Environmental Liability Insurance in China: Obstacles and Lessons

The prospects for mitigation of GHG emissions in China through liability insurance are not rosy. Because climate-change liability risk is a kind of environmental risk, the experience with environmental liability insurance in China is worth analysing for possible lessons and obstacles. Environmental tort liability cases are growing rapidly in China.[94] Meanwhile, environmental-risk insurance products are becoming increasingly available and are strongly promoted by the Chinese government.[95] Even if climate-change risk *were* covered under existing environmental policies, however, it is unlikely that such insurance could provide significant incentives for reduction of GHG emissions. The reasons are as follows.

First, the take-up rate for environmental liability insurance is low.[96] A glance at the environmental insurance market in China reveals three major insurance products covering environmental risks. The first product is a general liability insurance policy that was extended recently to cover several kinds of pollution risks.[97] The second is environmental liability insurance that mainly covers personal injuries and property damage caused by pollution. Chinese insurers started offering this stand-alone policy after 2007.[98] The third option involves pollution-site liability that provides coverage for damage to third parties as well as remediation costs of polluted sites.[99] Since 2007, eight provinces and cities (Jiangsu, Hubei, Hunan, Henan, Chongqing, Shenzhen, Ningbo, and Shenyang) have

---

[94] Daniel Carpenter-Gold, Note, *Castles Made of Sand: Public-interest Litigation and China's New Environmental Protection Law*, 39 Harvard Environmental Law Review 241 (2015).

[95] Michael G. Faure and Liu Jing, *Compensation for Environmental Damage in China: Theory and Practice*, 31 Pace Environmental Law Review 226 (2014).

[96] This opinion could be understood as that the penetration rate of environmental liability insurance is low.

[97] Michael G. Faure and Liu Jing, *Compensation for Environmental Damage in China: Theory and Practice*, 31 Pace Environmental Law Review 226 (2014).

[98] Michael G. Faure and Liu Jing, *Compensation for Environmental Damage in China: Theory and Practice*, 31 Pace Environmental Law Review 226 (2014).

[99] Michael G. Faure and Liu Jing, *Compensation for Environmental Damage in China: Theory and Practice*, 31 Pace Environmental Law Review 226 (2014).

been chosen to develop and promote liability insurance to cover various dangerous industrial risks, such as dangerous chemicals, dangerous waste treatment, and petrochemicals.[100] The revenue from environmental liability insurance only accounted for 0.015 percent of the total liability insurance revenue in these experimental areas in 2009.[101]

Second, obstacles to cover still exist under environmental liability insurance policies that concern pollution. *Pollution* is not clearly defined in the general environmental liability insurance policy.[102] Whether GHGs belong among pollutants remains an uncertainty. To avoid potential risk, insurers may modify policy language to exclude liability arising from climate-change claims against policyholders. When it comes to climate-related claims under environmental policies, third-party victims still need to prove a causal link between their losses and the insured's emissions.

As a result of the above debates concerning tortious liability rules, liability insurance coverage, and the environmental policies market in China, liability insurance would not be viable and might not be sufficient to encourage mitigation of GHG emissions at present. Government regulation may be the more efficient approach. China's approach to addressing GHG emissions relies heavily on "top-down, command-and-control regulation, built around bureaucratic targets and controls for local officials and state-owned enterprise leaders".[103] Most recently, the Chinese central government has made an international commitment to lower carbon dioxide emissions per unit of GDP by 60 to 65 percent of the 2005 level by 2030 and to start its national emissions trading

---

[100]   Lijing Liang and Jialin, [*The Ministry of Environmental Protection Tries to Promote Compulsory Environmental Liability Insurance*], Sina (July 9, 2012), available at http://green.sina.com.cn/2012-07-09/103024739500.shtml (China).

[101]   Lijing Liang and Jialin, [*The Ministry of Environmental Protection Tries to Promote Compulsory Environmental Liability Insurance*], Sina (July 9, 2012), available at http://green.sina.com.cn/2012-07-09/103024739500.shtml (China).

[102]   Michael G. Faure and Liu Jing, *Compensation for Environmental Damage in China: Theory and Practice*, 31 Pace Environmental Law Review 226, 268 (2014).

[103]   Alex Wang, *Climate Change Policy and Law in China* in Oxford Handbook of International Climate Change Law (Cinnamon P. Carlarne, Kevin R. Gray and Richard Tarasofsky eds., 2016). Examples of top-down, command-and-control regulation include but are not limited to: central authorities being given greater authority to punish bureaucrats and state-owned enterprise leaders for failing to meet energy, carbon, and pollution targets; central authorities issuing detailed implementation rules and offering training to assist local government actors in implementation; and central regulators developing multiple data sources and more readily observable proxies for target implementation. See Alex Wang, *The Search for Sustainable Legitimacy: Environmental Law and Bureaucracy in China*, 37 Harvard Environmental Law Review 365 (2013).

system in 2017.[104] In the future, the government could also establish a climate-change compensation fund and require major emitters to bear the responsibility for premiums.[105] The no-fault compensation mechanism of a climate-change fund could provide basic compensation for victims and could adopt the polluter-pays principle to encourage emitters to internalize the external cost of GHGs.[106] This scheme might be more viable than a tort-based liability insurance model since it could avoid the obstacles to tortious liability and liability insurance coverage.

V.   REGULATION BY CATASTROPHE INSURANCE FOR MITIGATION OF VALUE-AT-RISK

While third-party liability insurance might be infeasible for mitigating climate-change risks, catastrophe insurance has drawn attention as a means of mitigating climate-change risks—mainly value-at-risk—both in theoretical discussion and in practice.[107] In China, catastrophe insurance does have the potential to regulate insureds' behavior so as to mitigate risks.

A.   The Liability Insurance versus Catastrophe Insurance Models

Insurance is regarded as a powerful regulatory mechanism in many fields of liability, including automobile accidents, workplace injuries, environmental harms, corporate and securities liability, medical malpractice,

---

[104]   White House Office of the Press Secretary, US-China Joint Presidential Statement on Climate Change (September 25, 2015), available at https://www.whitehouse.gov/the-press-office/2015/09/25/us-china-joint-presidential-statement-climate-change.

[105]   Melissa Farris, *Compensation Climate Change Victims: The Climate Change Fund as an Alternative to Tort Litigation*, 2(2) Sea Grant Law and Policy Journal 49, 60 (2009–2010).

[106]   Melissa Farris, *Compensation Climate Change Victims: The Climate Change Fund as an Alternative to Tort Litigation*, 2(2) Sea Grant Law and Policy Journal 49, 60 (2009–2010).

[107]   For example, most government-sponsored catastrophe insurance programs are first-party insurance, such as the National Flood Insurance Program (NFIP) in the United States. See US Government Accountability Office, *National Flood Insurance Program: Greater Transparency and Oversight of Wind and Flood Damage Determinations Are Needed*, GAO-08-28 (2007). Some scholars discuss how to address catastrophic risks and emphasize the role of first-party insurance. See Michael Faure and Véronique Bruggeman, *Catastrophic Risks and First-Party Insurance*, 15 Connecticut Insurance Law Journal 1 (2008).

defective products, and defamation.[108] In the traditional framework, liability insurance can regulate potential tortfeasors to internalize the costs of the harms they create through its capacity to manage moral hazard and other bad acts of the insureds.[109] Thus, it improves the efficiency of risk mitigation. However, when it comes to climate-change risks, as discussed above, regulation by liability insurance to mitigate risks faces theoretical and practical obstacles.

In contrast to liability insurance, which focuses on GHG emitters, catastrophe insurance, which focuses mainly on victims of climate hazards, can overcome causal uncertainty and other tort liability obstacles. With respect to causal uncertainty, climate-change victims could seek compensation from catastrophe insurers directly without having to establish the liability of the emitters.[110] Catastrophe insurance shifts the causal connection between the emitters' activities and the victims' losses.

A catastrophe insurer has several regulatory techniques to minimize value-at-risk and protect insureds from climate hazards, including risk-based pricing, contract design, loss prevention services, and claim management.[111] Risk-based pricing is the most basic technique for creating incentives to reduce risk.[112] Risk-based premiums enable insurers to provide a discount to insureds adopting cost-effective mitigation measures and thus also to provide a clear signal to other insureds in the market. By including such elements as deductibles, copayments, coverage amount

---

[108]   Tom Baker and Rick Swedloff, *Regulation by Liability Insurance: From Auto to Lawyers Professional Liability*, 60 UCLA Law Review 1412 (2013); Tom Baker, *Liability Insurance as Tort Regulation: Six Ways that Liability Insurance Shapes Tort Law in Action*, 12 Connecticut Insurance Law Journal 1 (2005).

[109]   Omri Ben-Shahar and Kyle D. Logue, *Outsourcing Regulation: How Insurance Reduces Moral Hazard*, 111 Michigan Law Review 197, 210–212 (2012).

[110]   Hsin-Chun Wang, *Adaptation to Climate Change and Insurance Mechanism: A Feasible Proposal Based on a Catastrophe Insurance Model for Taiwan*, 9 NTU Law Review 317, 331 (2014).

[111]   These regulatory tools are widely discussed in series articles of regulation-by-insurance scholars. For example, Omri Ben-Shahar and Kyle D. Logue, *Outsourcing Regulation: How Insurance Reduces Moral Hazard*, 111 Michigan Law Review 197 (2012); Tom Baker and Rick Swedloff, *Regulation by Liability Insurance: From Auto to Lawyers Professional Liability*, 60 UCLA Law Review 1412 (2013); Shauhin Talesh, *Legal Intermediaries: How Insurance Companies Construct the Meaning of Compliance with Anti-Discrimination Laws*, 37 Law & Policy 209 (2015); Swenja Surminski, *The Role of Insurance in Reducing Direct Risk: The Case of Flood Insurance*, 7 International Review of Environmental and Resource Economics 241, 264 (2013).

[112]   Omri Ben-Shahar and Kyle D. Logue, *Outsourcing Regulation: How Insurance Reduces Moral Hazard*, 111 Michigan Law Review 197, 205–208 (2012).

limit, and exclusions, contract design can also be used to regulate risk both directly and indirectly. The deductible is an example. Deductibles can mitigate moral hazard directly because they prevent victims of climate hazards from shielding themselves entirely from loss and thus encourage them to exercise greater vigilance than would be the case without deductibles.[113] Furthermore, catastrophe insurers could provide loss-prevention services, such as flood-proofing of buildings or retrofitting of houses against windstorms.[114] With respect to climate-change risks, catastrophe insurers have worked in tandem with scientists to identify technical and economic parameters of climate-change risks and to develop system-wide technologies for loss prevention.[115] Claim management is necessary to control the ex post moral hazard of insureds, as a result of their inability to change the possibility of a climate hazard despite their ability to mitigate disaster losses.[116] Catastrophe insurers also employ adjusters to investigate claimed losses, measure them, and negotiate payouts to provide greater uniformity and predictability.

In terms of the catastrophe insurance model, drawbacks may also exist along with the above merits. First, because of the highly correlated nature and potential concentration of losses from extreme weather exposures, insurers' capacity and appetite to cover such losses may not be sufficient.[117] For this reason, at least in US Golf Coast states, standard homeowners' insurance policies typically exclude losses from floods and wind. Even endorsements that add coverage for these events at additional expense are normally unavailable.[118] The second drawback is that potential victims may fail to purchase insurance against low-probability extreme weather

---

[113]   Tom Baker and Rick Swedloff, *Regulation by Liability Insurance: From Auto to Lawyers Professional Liability*, 60 UCLA Law Review 1412 (2013).

[114]   Swenja Surminski, *The Role of Insurance in Reducing Direct Risk: The Case of Flood Insurance*, 7 International Review of Environmental and Resource Economics 241, 264 (2013).

[115]   P.A. Scott et al., *Human Contribution to the European Heat Wave of 2003*, 432 Nature 610 (2004).

[116]   This is workable because a policyholder is not concerned with the cost of a claim once it occurs. See Tom Baker and Rick Swedloff, *Regulation by Liability Insurance: From Auto to Lawyers Professional Liability*, 60 UCLA Law Review 1412, 1421 (2013).

[117]   Howard Kunreuther, *The Role of Insurance in Reducing Losses from Extreme Events: The Need for Public-Private Partnerships*, 40 Geneva Risk and Insurance Review 741 (2015).

[118]   Peter Molk, *Private Versus Public Insurance for Natural Hazards: Individual Behavior's Role in Loss Mitigation* in Risk Analysis of Natural Hazards 265 (Paolo Gardoni et al. eds., 2015). Endorsements are clauses in an insurance policy detailing an exemption from or change in coverage. When an endorsement modifies,

disasters because of the observed behavioral anomaly of individuals to underestimate the expected costs of extreme weather hazards and also because of repeated government bailouts.[119] Accordingly, these drawbacks associated with catastrophe insurance would need to be addressed to develop a feasible catastrophe insurance model.

B.   Developing a Feasible Catastrophe Insurance-based Private–Public Model for China

Catastrophe risks have never been covered systematically by either private insurance or government insurance in China. Before exploring a feasible mechanism for China, it is worth describing and assessing the current insurance framework related to the financial impact caused by climate hazards. There are two kinds of insurance systems in China: social insurance protection and the private insurance market. With regard to social insurance protection, Basic Medical Insurance and other compensation programs such as Occupational Injury Insurance provide basic financial protection.[120] These programs provide basic social security protection to mitigate financial losses caused by climate-related disasters. The private insurance market in China consists of life insurance (including private health insurance and short-term disability insurance) and property-casualty insurance.[121] Although China's insurance market ranked as the fourth largest in the world in 2013,[122] insurance that offers coverage for damages caused by natural disasters is undeveloped in China. For example, following the 2008 Great Sichuan Earthquake, only 0.3 percent of the total

---

qualifies, or restricts the terms of the original policy, the endorsement controls. See Steven Plitt et al., 2 Couch on Insurance § 21:22 (3rd edn, 2011).

[119]   H. Kunreuther, M.V. Pauly and S. McMorrow, Insurance and Behavioral Economics: Improving Decisions in the Most Misunderstood Industry 114–115 (2013).

[120]   Social insurance consists of five parts: basic old-age insurance, basic medical insurance, occupational injury insurance, unemployment insurance, and maternity insurance. The five components of the social insurance program and a public reserve fund for housing are together described as the "Five Insurances and One Fund" and provide basic protection for the Chinese people. See Bingzheng Chen, Sharon Tennyson, Maoqi Wang and Haizhen Zhou, *The Development and Regulation of China's Insurance Market: History and Perspectives*, 17 Risk Management and Insurance Review 241, 243 (2014).

[121]   Bingzheng Chen, Sharon Tennyson, Maoqi Wang and Haizhen Zhou, *The Development and Regulation of China's Insurance Market: History and Perspectives*, 17 Risk Management and Insurance Review 241, 243 (2014).

[122]   Xinhua News Agency (2013), available at http://www.gov.cn/xinwen/2014-07/09/content_2714415.htm.

losses were covered by insurance companies.[123] It was only in 2014 that the central government trialed the catastrophe insurance program in the cities of Shenzhen and Ningbo, and the region of Chuxiong where the catastrophe risk would be shared between the government, insurance and reinsurance firms, and individuals.[124] In considering the current insurance market in China, a phased catastrophe insurance-based private–public model that marries the merits of both insurance and government in the mitigation of value-at-risk might address the problems described above. The nuts and bolts of this mechanism are as follows.

First, under such a model, the state would require all property insurers to provide catastrophe coverage for residential properties. As we discussed in Chapter 2, mandatory insurance could help to solve supply and demand anomalies and thus improve efficiency, correct market failure, promote distributive justice, price accurately, and regulate more efficiently.[125]

In doing so, it would be important for the government to avoid interfering with private insurers' actuarially fair pricing. Insurance premiums reflecting risks can provide individuals with accurate signals in cost-effective mitigation measures.[126] This may be difficult because the Chinese government could face political pressure to interfere with insurers' risk classification and risk-based premiums, particularly because risk classification is not always compatible with social solidarity objectives in China, which promote equal treatment of all citizens.[127] Nevertheless, even if there are concerns about the affordability of catastrophe insurance, it is better for the government to provide subsidies to lower-income residents (such as through insurance vouchers) than to suppress insurers' incentives to underwrite policies and distort risk signals provided by actuarially based premiums. Therefore, the government should neither create new institutions to supplant private solutions nor suppress premiums of insurance policies. As a Chinese proverb says, "You can't expect the horse to

---

[123] *Establishing Catastrophe Insurance System faces acceleration*, China Youth Daily (March 14, 2011), available at http://zqb.cyol.com/html/2011-03/14/nw.D110000zgqnb_20110314_1-05.htm?div=-1.

[124] *China says testing catastrophe insurance system*, Reuters (August 20, 2014), available at http://www.businessinsurance.com/article/20140820/NEWS04/140829 990?AllowView=VDl3UXk1T3hDUFNCbkJiYkY1TDJaRUt0ajBRV0ErOVVH UT09#.

[125] The reasons have been discussed in Chapter 2, section IV.B.2.

[126] Howard Kunreuther, *The Role of Insurance in Reducing Losses from Extreme Events: The Need for Public-Private Partnerships*, 40 Geneva Risk and Insurance Review 741 (2015).

[127] J.D. Cummins and O. Mahul, *Catastrophe Risk Financing in Developing Countries: Principles for Public Intervention*, World Bank Publications 76 (2009).

run fast when you don't let it graze". The state cannot expect insurers to underwrite policyholders' risks if it does not let insurers make profits. Furthermore, that type of rate regulation would undermine the effectiveness of insurance as a risk mitigation device.

Second, recognizing private insurers' reluctance to underwrite catastrophe risks without any meaningful underwriting history, predictive model, or sufficient capacity, the state could provide a government-funded backstop that assumes the risks of any losses above a predetermined threshold, thereby capping private insurers' maximum exposure. This type of cap could promote the initial entry of private insurers.[128] The coastal states of the United States have established many quasi-public residual risk insurance programs for wind exposures when private insurers have dropped or threaten to withdraw wind polices.[129] For example, the Florida Citizens Property Insurance Corporation, the Louisiana Citizens Property Insurance Association, the Texas Windstorm Insurance Association, the Mississippi Wind Pool and the North Carolina Coastal Wind Pool have all been created as last resort to cover excess risks.[130]

Third, after several years (maybe longer), when private insurers have gained underwriting experience and developed models, this type of government support could be removed. However, this may be easier said than done. In this respect, the US terrorism coverage program, Terrorism Risk Insurance Act (TRIA), provides a cautionary lesson. Established after the terrorist attacks of September 11, 2001, TRIA has been twice extended for a total of nine years past its original 2005 horizon.[131] This is a good

---

[128] For example, following Hurricane Andrew, the Florida Hurricane Catastrophe Fund, established to reimburse all insurers for a portion of their losses from catastrophic hurricanes, has proven that a government reinsurance facility can assist catastrophe coverage when insurers are inclined to withdraw from the catastrophe insurance market. See Howard Kunreuther, *The Role of Insurance in Reducing Losses from Extreme Events: The Need for Public-Private Partnerships*, 40 Geneva Risk and Insurance Review 741 (2015).

[129] Donald T. Hornstein, *Lessons From U.S. Coastal Wind Pools: About Climate Finance and Politics*, 43 Boston College Environmental Affairs Law Review 345 (2016).

[130] Donald T. Hornstein, *Lessons From U.S. Coastal Wind Pools: About Climate Finance and Politics*, 43 Boston College Environmental Affairs Law Review 345 (2016).

[131] TRIA, which has been extended for an additional six years, shows the practical difficulties in committing to a "temporary" subsidy. See Peter Molk, *Private Versus Public Insurance for Natural Hazards: Individual Behavior's Role in Loss Mitigation* in Risk Analysis of Natural Hazards 265 (Paolo Gardoni et al. eds., 2015).

example of why government support is difficult to phase out over time. One possible reason is that insurers have lobbied to keep the Act in place.

Fourth, the state can use its powers of compulsion to bring more residents into the pool, which maximizes risk spreading, in turn maximizing consumer participation in the insurance market and minimizing adverse selection concerns. Compulsory insurance is practiced in some countries, such as in the Caisse Centrale de Réassurance (CCR) program in France[132] and the Turkish Catastrophe Insurance Pool (TCIP).[133] An insurance mandate is especially important in China, given its recent history of a planned economy.[134] The Chinese people still have a strong reliance on government, especially in the aftermath of a catastrophe. Under the Whole-Nation System, the government is committed to bailing victims out after a disaster. According to an empirical study on property and casualty insurance in five Chinese provinces, there is a negative correlation between the amount of government relief and residents' investment in prevention measures such as purchasing insurance.[135] Without a compulsory requirement, residents in China will seldom purchase insurance to transfer casualty risks.

Fifth, the state should give special treatment to low-income individuals residing in hazard-prone areas because of equity and affordability issues.[136] For example, the state could provide means-tested vouchers to maintain risk-based premiums while covering part of the cost of insurance.[137] As a

---

[132]   The CCR, as a mandatory comprehensive disaster insurance model, has existed in France since 1987 and has also been recently introduced in Belgium and is under consideration in other countries. See G. Dari-Mattiacci and M.G. Faure, *The Economics of Disaster Relief*, 37 Law & Policy 180 (2015).

[133]   The TCIP is regarded as one of the best examples of private–public partnerships in an emerging market designed to reduce economic losses from disasters. See B. Burcak Başbuğ-Erkan and Ozlem Yilmaz, *Successes and Failures of Compulsory Risk Mitigation: Re-evaluating the Turkish Catastrophe Insurance Pool*, 39 Disasters 782 (2015).

[134]   Under the planned economic system, private insurance markets would cease to exist because the government would bear all risks and cover individuals' exposures.

[135]   L. Tian and Y. Zhang, *Influence Factors of Catastrophe Insurance Demand in China–Panel Analysis in a Case of Insurance Premium Income of Five Provinces [Woguo Juzai Baoxian Xuqiu Yingxiang Yinsu Shizheng Yanjiu: Jiyu Wusheng Bufen Baofei Shouru Mianban Yanjiu]*, 26 Wuhan University of Technology (Social Science Edition) [Wuhan Ligong Daxue Xuebao (Shehui Kexue Ban)] 175 (2013).

[136]   Under a compulsory insurance program, low-income individuals may struggle to afford payments.

[137]   Howard Kunreuther, *The Role of Insurance in Reducing Losses from Extreme Events: The Need for Public-Private Partnerships*, 40 Geneva Risk and Insurance Review 741 (2015).

condition for providing the voucher, the government could then require property owners to invest in mitigation based on their affordability.

Sixth, besides intervention in the insurance market, the state could adopt administrative measures or regulations to require residents to take mitigation measures. Such measures or regulations should coordinate with risk-based insurance premiums. For example, the state could update its building code standards and enforce high-wind design provisions for residential housing.

VI.   CONCLUSION

In the context of climate-change risk management, either private insurance or the state can play a crucial role in mitigating risks. As the preceding discussion of liability insurance and catastrophe insurance makes clear, catastrophe insurance is a much more feasible method of regulating climate-change risks. A compulsory catastrophe insurance-based private–public model will not only enhance mitigation of value-at-risk but also provide victims with sufficient financial protection against climate hazards that are not excluded. This hybrid mechanism has become a prototype for developing catastrophe insurance in several countries. It should be developed as soon as possible in China to cope with the increasing risks of climate change.

4. Regulation by catastrophe insurance: a comparative study[1]

I. INTRODUCTION

According to Beck, the current era is characterized as a "risk society".[2] Under the influence of climate extremes and other natural disasters, the world is exposed to more and more catastrophe risks.[3] Although a catastrophe risk event occurs infrequently, it causes significant human and financial losses. There is increasing attention not only to the question how to compensate victims, but also to how compensation mechanisms, including insurance, can stimulate disaster risk reduction.[4]

Increasingly, insurance is seen as a tool to "outsource" public regulation.[5] In order to remedy the risk of moral hazard inherent in any insurance contract, insurers "regulate" how organizations and individuals should

---

[1] The previous version of this chapter appears in Qiaho He and Michael Faure, *Regulation by Catastrophe Insurance: A Comparative Study*, 24 Connecticut Insurance Law Journal 89 (2018).

[2] Ulrich Beck, Risk Society: Towards a New Modernity (1992).

[3] Data from large reinsurers show that the amounts and damage resulting from both man-made and natural disasters have been increasing over the past 30 years. See Swiss Re, Sigma, *Natural Catastrophes and Man-made Disasters in 2014: Convective and Winter Storms Generate Most Losses* (2015), available at http://www.actuarialpost.co.uk/downloads/cat_1/sigma2_2015_en.pdf; Munich Re, *Topics GEO 2013* (2013), available at http://www.munichre.com/site/corporate/get/documents_E1043212252/mr/assetpool.shared/Documents/5_Touch/_Publications/302-08121_en.pdf.

[4] For example, The 2005 Hyogo Framework for Action highlights the urgency to advance the expansion of insurance markets to finance risk following a natural disaster. See J.D. Cummins and O. Mahul, Catastrophe Risk Financing in Developing Countries xi (2009). In the EU the Green Paper on the insurance of natural and man-made disasters equally pays attention to the ability of insurance to provide compensation and to stimulate risk-mitigating behavior. See European Commission, Green Paper on the insurance of natural and man-made disasters, COM 213 final (2013).

[5] Omri Ben-Shahar and Kyle D. Logue, *Outsourcing Regulation: How Insurance Reduces Moral Hazard*, 111 Michigan Law Review 197 (2012).

deal with specific risks.[6] Private insurance can act not only as a form of post-disaster relief but also as a form of private regulation—a contractual device controlling and motivating behavior prior to the occurrence of losses.[7] Insurance is a tool of risk management that addresses three aspects of risk management: risk assessment (or risk analysis), risk control, and risk financing.[8] From society's perspective insurance has at least two important functions. The first is that it can spread risks over a larger community and thus compensate risk-averse individuals exposed to risky activities through risk pooling and risk shifting. A second function is that by controlling the moral hazard risk insurers also regulate policyholders' behavior and can thus contribute to risk reduction. Insurance may have these important functions also in relation to catastrophic risks, provided specific conditions are met.[9] As more and more catastrophic losses are caused by a growing movement of population to high-risk areas, such as the coastal area in Florida, by aging infrastructure, and by low levels of public and private investment in risk reduction measures, insurance has an increasingly important potential to mitigate risk and prevent loss through regulating risky behaviors.[10] Once insurers underwrite catastrophe risk, they have every reason to work to reduce their payouts. Therefore, regulation by insurance may help in realizing the goal of disaster risk reduction and the corresponding losses.

Although there is general agreement on the important contribution of insurers to disaster risk reduction, less is known about the precise instruments and techniques used by insurers to achieve this. This chapter identifies under which specific conditions insurance can function as a substitute for, or a complement to, government regulation of catastrophic risks associated with natural disasters. The chapter identifies five regulatory techniques of catastrophe insurance that may complement, and in some cases perhaps even outperform, government regulation by creating incentives for optimal behavior for individuals and organizations. The

---

[6]   Steven Shavell, *On Moral Hazard and Insurance*, 93 Quarterly Journal of Economics 541 (1979).

[7]   Omri Ben-Shahar and Kyle Logue, *The Perverse Effects of Subsidized Weather Insurance*, 68 Stanford Law Review 571 (2016).

[8]   F. Outreville, Theory and Practice of Insurance 45–64 (1998); R. Thoyts, Insurance Theory and Practice 286–295 (2010); Emmett J. Vaughan and Therese M. Vaughan, Fundamentals of Risk and Insurance 16 (2007).

[9]   George L. Priest, *The Government, the Market, and the Problem of Catastrophe Loss*, 12 Journal of Risk and Uncertainty 219 (1996).

[10]   Erwann Michel-Kerjan, *Have We Entered an Ever-Growing Cycle on Government Disaster Relief*, Presentation to US Senate Committee on Small Business and Entrepreneurship (2013).

chapter then compares five middle- to high-income countries—the United Kingdom, the United States, France, Japan, and Turkey—in which catastrophic risks is regulated by insurance. In the comparison I analyse the role of the state in facilitating regulation by insurance by examining to what extent the tools to control moral hazard are encouraged or restricted by government regulation.

Section II reviews the literature describing how, also in the field of catastrophe insurance, insurers can exercise regulatory functions, aiming at disaster risk reduction. Section III discusses five specific tools that can contribute to disaster risk reduction. Section IV provides examples of the use of those regulatory techniques in five different countries, both developed and developing. Section V provides a comparative discussion concerning the effectiveness of regulation via catastrophe insurance. Section VI discusses the possibility and feasibility of regulation by catastrophe insurance in China, where it is not yet utilized; and section VII concludes.

II. INSURANCE AS A TOOL OF DISASTER RISK REDUCTION

Any insurance contract, whether it is a first party (victim) insurance or a third party (liability) insurance, is vulnerable to the moral hazard risk. For many, insurance is, precisely as a result of moral hazard, the antipode of risk reduction.[11] Moral hazard is the tendency of insureds from vulnerable areas to exercise less care to avoid losses than they would if the losses were not covered by insurers.[12] It is, one must admit, only logical for the insureds to change their behavior as soon as the risk is fully removed from them.[13] Changing that behavior is in that sense not "immoral". However, if moral hazard cannot be remedied, liability insurance may have socially negative consequences as it could dilute the incentives of the potential tortfeasor to invest in prevention. In that case liability insurance should be prohibited since it would increase risk in society.[14] Remedies for moral hazard are available. A first best option is a monitoring by the insurer

---

[11]   Kenneth Arrow, Essays in the Theory of Risk-Bearing (1971); Bengt Hölmstrom, *Moral Hazard and Observability*, 10 Bell Journal of Economics 74 (1979).

[12]   Kenneth S. Abraham, Insurance Law and Regulation 7 (5th edn, 2010).

[13]   Mark Pauly, *The Economics of Moral Hazard: Comment*, 58 The American Economic Review 531 (1968).

[14]   Steven Shavell, *On Liability and Insurance*, 13 Bell Journal of Economics 120 (1982).

and a corresponding adaptation of premium conditions.[15] This solution is referred to as "first best" since risk would be completely removed from a risk-adverse individual. A second best solution is still to impose a partially through risk, by applying deductibles or an upper limit on cover.[16] In practice insurers will apply a combination of different techniques (risk differentiation, specific conditions in policies and deductibles) to control moral hazard.[17] It is precisely through this control of the moral hazard risk that insurers will act as de facto regulators and invest in risk prevention. I will show that through this control of the moral hazard risk insurers are viewed as private (risk) regulators (A) and that precisely through this control of moral hazard they can equally contribute to disaster risk reduction (B).

A.   Insurance as Private (Risk) Regulation

Regulation by insurance is not the same as insurance regulation. The latter is a classic topic of insurance law, and mainly discusses how the insurance business and organizations are regulated by administrative agencies.[18] On the other hand, regulation by insurance explores the potential value of insurance as a complement to, or substitute for, the State.[19] There is an increasing interest in the regulatory potential of insurance companies both in academic literature as well as at the policy level.

A considerable amount of literature has been devoted to discussing

---

[15]   This can take place ex ante (through a so-called risk classification) or ex post (after the accident) through experience rating. The latter implies effectively that the premium would be increased after a reported incident.

[16]   Hsin-Chun Wang, *Adaptation to Climate Change and Insurance Mechanism: A Feasible Proposal Based on a Catastrophe Insurance Model for Taiwan*, 9 NTU Law Review 317 (2014).

[17]   Steven Shavell, *On Moral Hazard and Insurance*, 93 Quarterly Journal of Economics 541 (1979); Steven Shavell, *On the Social Function and Regulation of Liability Insurance*, 25 The Geneva Papers on Risk and Insurance—Issues and Practice 166 (2000).

[18]   There are many possible ways to describe insurance regulation. For example, the function of insurance regulation describes seven main functional divisions: licensing, taxation, solvency, rates, forms, access and availability, and market conduct; theoretical justifications of insurance regulation present information problems, externalities, opportunism and egalitarian or distributional objectives to justify regulation. See Tom Baker and Kyle D. Logue, Insurance Law and Policy: Cases and Materials 573–580 (3rd edn, 2013).

[19]   Some scholars prefer the term "governance by insurance". In this chapter, "governance by insurance" is interchangeable with the term "regulation by insurance".

regulation by insurance. As far back as 1986, Reichman explored insurance as a social control tool to regulate crime risk.[20] More recently, Abraham offered an overview and critique of modern conceptions of insurance based on the debates about insurance and insurance law in recent decades, one of which is the conception of "insurance as governance", which corresponds to the idea of regulation by insurance.[21]

In 2002, Heimer discussed the cost and benefit of private regulation through insurance.[22] In 2003, Ericson, Doyle, and Barry adopted a sociological perspective to explore insurance as governance, and documented how the insurance industry governs our lives and asserts insurance governing through nine interconnected dimensions.[23] In 2004, Ericson and Doyle further applied their theoretical framework to four sets of risks that are governed by insurance: life, disability, earthquakes, and terrorism.[24] They invented a new term for the insurance approach to natural catastrophic risk: absorbing risk, which requires creating an infrastructure that can withstand the catastrophe.[25] Consistent with the concept of absorbing risk, Baker urged the reconsideration of assessment approaches to catastrophe insurance to allow insurance institutions to manage the uncertainties of catastrophic risk.[26]

In 2005, Baker and Farrish initiated the discussion of the technique of firearms regulation by liability insurance.[27] With gun violence dominating the headlines over the last few years, Kochenburger also argues that

---

[20]   Nancy Reichman, *Managing Crime Risks: Toward an Insurance Based Model of Social Control* in Research in Law, Deviance and Social Control 151–172 (Andrew T. Scull and Stephen Spitzer eds.,1986).

[21]   K.S. Abraham, *Four Conceptions of Insurance*, 161 University of Pennsylvania Law Review 653 (2013). See also J.W. Stemple, *The Insurance Policy as Social Instrument and Social Institution*, 51 William & Mary Law Review 1489 (2009). Stemple's argument seems to fit comfortably within the insurance as governance conception.

[22]   Carol Heimer, *Insuring More, Ensuring Less: The Costs and Benefits of Private Regulation through Insurance* in Embracing Risk: The Changing Culture of Insurance and Responsibility 117–145 (T. Baker and J. Simon eds., 2002).

[23]   R.V. Ericson, A. Doyle and D. Barry, Insurance as Governance (2003).

[24]   R.V. Ericson and A. Doyle, Uncertain Business: Risk, Insurance and the Limits of Knowledge (2004).

[25]   R.V. Ericson and A. Doyle, Uncertain Business: Risk, Insurance and the Limits of Knowledge 34 (2004).

[26]   Tom Baker, *Embracing Risk, Sharing Responsibility*, 56 Drake Law Review 561 (2007).

[27]   Tom Baker and Thomas O. Farrish, *Liability Insurance and the Regulation of Firearms* in Suing the Gun Industry: A Battle at the Crossroads of Gun Control and Mass Torts 292 (Timothy D. Lytton ed., 2005).

regulation by liability insurance could serve as a potentially valuable tool to address and reduce gun violence;[28] however, Mocsary contends that the insurance regime is likely to attain this goal.[29] Besides gun violence, Yin, Kunreuther, and White examine how environmental liability insurance can reduce environmental accidents based on disaggregated (facility-level) data.[30] Ben-Shahar and Logue explore regulation by insurance as a substitute for government regulation of safety in areas of products liability, workers' compensation, auto and homeowners', environmental liability and tax liability, and expand to yet unutilized areas, such as consumer protection, food safety, and financial statements.[31] Baker and Swedloff summarize regulation by liability insurance, and draw upon prior literature to examine four areas of liability and corresponding insurance—shareholder liability, automobile liability, gun liability, and medical professional liability—and develop a conceptual framework to guide qualitative research for lawyers' professional liability.[32] In 2015, Talesh significantly widened the scope of regulation by insurance through the study of Employment Practices Liability Insurance (EPLI), explaining how insurance practices transform the moral logic of anti-discrimination law into the risk management logic of EPLI loss prevention advice. This study shows how regulation by insurance does not simply consist of assessing how well liability insurance delivers a legal deterrence signal, but also how it transforms that signal into loss prevention.[33] In 2017, Rappaport described and assessed the contemporary market for liability insurance in the policing context and the effects of insurance on police behavior.[34]

As well as insurers playing a role as private regulators, recently it has also been stressed that reinsurance companies can act as "silent

---

[28]   Peter Kochenburger, *Liability Insurance and Gun Violence*, 46 Connecticut Law Review 1265 (2014).

[29]   George Mocsary, *Insuring Against Guns?*, 46 Connecticut Law Review 1209 (2014).

[30]   Haitao Yin, Howard Kunreuther and Matthew White, *Risk-Based Pricing and Risk-Reducing Effort: Does the Private Insurance Market Reduce Environmental Accidents?*, 54 Journal of Law and Economics 325 (2011).

[31]   Omri Ben-Shahar and Kyle D. Logue, *Outsourcing Regulation: How Insurance Reduces Moral Hazard*, 111 Michigan Law Review 197 (2012).

[32]   Tom Baker and Rick Swedloff, *Regulation by Liability Insurance: From Auto to Lawyers Professional Liability*, 60 UCLA Law Review 1412 (2013).

[33]   Shauhin Talesh, *Legal Intermediaries: How Insurance Companies Construct the Meaning of Compliance with Anti-Discrimination Laws*, 37 Law & Policy 209 (2015).

[34]   John Rappaport, *How Private Insurers Regulate Public Police*, 130 Harvard Law Review 1539 (2017).

regulators",[35] more particularly in exercising a regulatory influence on insurers.[36] As I will show below, there are often also hybrid constructions where catastrophe insurance is offered by reinsurance pools in which the government equally participates.

B.   Disaster Risk Reduction by Controlling Moral Hazard

The regulatory effects of insurance can also be found in the area of catastrophe insurance. Starting from the danger of moral hazard, the control exercised by insurers (to remedy moral hazard) will also provide effective incentives for disaster risk reduction in the field of catastrophes. Kunreuther and his colleagues at the Wharton Risk Management and Decision Processes Center argue that insurance can be structured to improve the incentive of people to protect themselves against flood and hurricane damage. To achieve this goal, they propose multi-year insurance contracts with risk-based premiums that could enable insurers to lower premiums for properties where measures have been taken to reduce risk.[37] As for climate-related extremes, Telesetsky contends that third-party insurance that follows the polluter pays principle could compel timely climate change mitigation on the part of major greenhouse gas emitters.[38] Telesetsky also explores how mandatory climate change catastrophe insurance can serve the goals of both corrective and distributive justice. Furthermore, Faure and Bruggeman, from the perspective of compensation, document how first-party insurance can constitute a viable alternative to government compensation while victims also benefit from preventative incentives.[39]

---

[35]   Aviva Abramovsky, *Reinsurance: The Silent Regulator?*, 15 Connecticut Insurance Law Journal 345 (2009).
[36]   Marcos Antonio Mendoza, *Reinsurance as Governance: Governmental Risk Managements Pools as a Case Study in the Governance Role Played by Reinsurance Institutions*, 21 Connecticut Insurance Law Journal 53 (2014).
[37]   Howard Kunreuther, *The Role of Insurance in Reducing Losses from Extreme Events: The Need for Public–Private Partnerships*, 40 The Geneva Risk and Insurance Review 741 (2015); Howard Kunreuther and Erwann Michel-Kerjan, *Managing Catastrophic Risks through Redesigned Insurance: Challenges and Opportunities* in Handbook of Insurance 517 (G. Dionne ed., 2013); Howard Kunreuther and Mark Pauly, *Insuring Against Catastrophes* in The Known, the Unknown and the Unknowable in Financial Risk Management 210 (F.X. Diebold, N. Doherty and R. Herring eds., 2010).
[38]   Anastasia Telesetsky, *Insurance as a Mitigation Mechanism: Managing International Greenhouse Gas Emissions through Nationwide Mandatory Climate Change Catastrophe Insurance*, 27 Pace Environmental Law Review 691 (2010).
[39]   Michael G. Faure, *Insurability of Damage Caused by Climate Change: A Commentary*, 155 University of Pennsylvania Law Review 1875 (2007); Michael

Empirical evidence supports this literature. An empirical study of catastrophe insurance markets in Germany and the United States utilizing field survey data suggests the opposite of a moral hazard effect.[40] This study responds to the theoretical hypothesis that recognizes that insurers have the capacity and means to manage moral hazard.[41] The findings from Germany conclude that "individuals with flood insurance are more likely to have undertaken one of the suggested flood coping measures than uninsured households".[42] This conclusion is supported by the evidence from the United States, which shows that households that are more likely to have flood insurance and homeowners' policies that cover wind damage engage in more ex ante property risk reduction behavior on hurricane preparedness.[43] There is also evidence from Switzerland, where a public monopoly insurance exists with mandatory participation, that the insurance scheme provides significant incentives for risk reduction.[44]

Catastrophic risk may result in significant human and financial losses, and is therefore an issue that the State must address. The State cooperates with the insurance industry to regulate and absorb some of the catastrophe risk. The State needs the cooperation of the insurance industry because of the low-frequency but high-impact nature of catastrophe risk, and

---

Faure and Véronique Bruggeman, *Catastrophic Risks and First-Party Insurance*, 15 Connecticut Insurance Law Journal 1 (2008).

[40]   The authors conducted a comprehensive empirical study of risk selection in natural disaster insurance markets, and asked whether disaster preparedness activities differ when people have natural disaster coverage. The statistical analyses are based on survey data of individual disaster insurance purchases and risk mitigation activities in Germany and the United States. See Paul Hudson et al., *Moral Hazard in Natural Disaster Insurance Markets: Empirical Evidence from Germany and the United States*, 93 Land Economics 179 (2017).

[41]   E.g. Steven Shavell, *On Moral Hazard and Insurance*, 93 Quarterly Journal of Economics 541 (1979); Omri Ben-Shahar and Kyle D. Logue, *Outsourcing Regulation: How Insurance Reduces Moral Hazard*, 111 Michigan Law Review 197 (2012).

[42]   Paul Hudson et al., *Moral Hazard in Natural Disaster Insurance Markets: Empirical Evidence from Germany and the United States*, 93 Land Economics 179 (2017).

[43]   Paul Hudson et al., *Moral Hazard in Natural Disaster Insurance Markets: Empirical Evidence from Germany and the United States*, 93 Land Economics 179 (2017).

[44]   P. Raschky et al., Alternative Financing and Insurance Solutions for Natural Hazards: A Comparison of Different Risk Transfer Systems in Three Countries—Germany, Austria and Switzerland—Affected by the August 2005 Floods 13–17 (2009); G. Kirchgässner, *On the Efficiency of a Public Insurance Monopoly: The Case of Housing Insurance in Switzerland* in Public Economics and Public Choice 221 (P. Baake and R. Bork eds., 2007).

the complexity of establishing affordable and sustainable management and compensation arrangements.[45] Moreover, the State often creates regulatory vacuum by refusing to take up contentious questions in which activities related to catastrophe risk should be encouraged, permitted, or proscribed.[46] The insurance industry can address the problems caused by catastrophe risk and fill the regulatory vacuum by providing the technical apparatus needed for risk reduction and loss compensation.[47] In turn, the State cooperates with the insurance industry in relation to catastrophic losses which may exceed amounts that could be insured on normal insurance and reinsurance markets. In such cases the State provides compensation of an upper layer as reinsurer of last resort.[48] In many legal systems there are various mutual dependencies between the State and the insurance industry in the cover of natural disasters: the State depends upon the insurance industry to provide primary cover and incentives for disaster risk reduction. Insurers, on the other hand, also rely on primary investments by the State in disaster risk reduction (for example building dikes and levees) and regulating disaster risk reduction. Insurers, moreover, also depend upon the State as reinsurer of last resort to provide an upper layer of cover.[49] Moreover, in some cases reinsurance is provided by pools that have a hybrid character as they may involve both reinsurers and the government.[50] The question, however, arises as to what

---

[45]   Youbaraj Paudel, *A Comparative Study of Public–Private Catastrophe Insurance Systems: Lessons from Current Practices*, 37 The Geneva Papers on Risk and Insurance—Issues and Practice 257 (2012).

[46]   Carol Heimer, *Insuring More, Ensuring Less: The Costs and Benefits of Private Regulation through Insurance* in Embracing Risk: The Changing Culture of Insurance and Responsibility 117 (T. Baker and J. Simon eds., 2002).

[47]   Tom Baker, *Insurance in Sociolegal Research*, 6 Annual Review of Law and Social Science 433 (2010).

[48]   Johanna Hjalmarsson and Mateusz Bek, *Legislative and Regulatory Methodology and Approach: Developing Catastrophe Insurance in China* in Insurance Law in China 191 (Johanna Hjalmarsson and Dingjing Huang eds., 2015); Joanne Linnerooth-Bayer et al., *Insurance against Losses from Natural Disasters in Developing Countries: Evidence, Gaps and the Way Forward*, 1 Journal of Integrated Disaster Risk Management 1 (2011); Véronique Bruggeman, Michael G. Faure and Karine Fiore, *The Government as Reinsurer of Catastrophe Risks?*, 35 The Geneva Papers on Risk and Insurance—Issues and Practice 369 (2010).

[49]   Veronique Bruggeman, Michael Faure and Tobias Heldt, *Insurance Against Catastrophe: Government Stimulation of Insurance Markets for Catastrophic Events*, 23 Duke Environmental Law and Policy Forum 185 (2012).

[50]   Marcos Antonio Mendoza, *Reinsurance as Governance: Governmental Risk Managements Pools as a Case Study in the Governance Role Played by Reinsurance Institutions*, 21 Connecticut Insurance Law Journal 53 (2014).

are the precise technical tools employed by insurers to aim at disaster risk reduction and to what extent is the effectiveness of those tools dependent upon their cooperation with the state?

III. REGULATORY TECHNIQUES OF CATASTROPHE INSURANCE

As a private regulator, insurance operates stealthily by using technical tools to reduce moral hazard.[51] As indicated, instruments to control moral hazard on the one hand consist of techniques to control the behavior of the insured via adapted policy conditions, and on the other hand consist in partially exposing the insured to risk.[52] In the literature a further refinement of the regulatory techniques of insurance has been made leading to the following taxonomy:[53] these technical tools, which almost all insurers use to one degree or another, include risk-based pricing, contract design (e.g. limits, deductibles, copayments, and exclusions), loss prevention, claim management, and refusal to insure. Of course, not all of those technical tools will be used by catastrophe insurers to the same extent. However, this taxonomy provides a good categorization of the types of technical tools usually employed in catastrophe insurance to control moral hazard.

---

[51]   Carol Heimer, *Insuring More, Ensuring Less: The Costs and Benefits of Private Regulation through Insurance* in Embracing Risk: The Changing Culture of Insurance and Responsibility 117 (T. Baker and J. Simon eds., 2002).

[52]   Steven Shavell, *On Moral Hazard and Insurance*, 93 Quarterly Journal of Economics 541 (1979).

[53]   Tom Baker and Thomas Farrish developed taxonomy of the various types of "regulation by liability insurance". Victor Goldberg further illustrated some of these techniques in the area of liability insurance. Omri Ben-Shahar and Kyle D. Logue relied on those prior taxonomies and highlighted the advantages that insurance has relative to government regulation. Tom Baker and Rick Swedloff discerned these taxonomies in different liability areas for their ongoing research on lawyers' professional liability insurance. See Tom Baker and Thomas O. Farrish, *Liability Insurance and the Regulation of Firearms* in Suing the Gun Industry: A Battle at the Crossroads of Gun Control and Mass Torts 292 (Timothy D. Lytton ed., 2005); Tom Baker and Rick Swedloff, *Regulation by Liability Insurance: From Auto to Lawyers Professional Liability*, 60 UCLA Law Review 1412 (2013); Omri Ben-Shahar and Kyle D. Logue, *Outsourcing Regulation: How Insurance Reduces Moral Hazard*, 111 Michigan Law Review 197 (2012); Victor P. Goldberg, *The Devil Made Me Do It: The Corporate Purchase of Insurance*, 5 Review of Law and Economics 541 (2009).

A. Risk-based Pricing

Risk-based pricing is considered to be the most basic technique for creating incentives to reduce risk.[54] Insurers set premiums to reflect underlying risk levels in order to provide individuals with incentives to mitigate losses.[55] Indeed, insurers often adopt feature rating[56] and experience rating in order to signal premium loss prevention. Charging lower premiums to careful policyholders induces them to reduce exposure to claims in order to avoid higher premiums in the future.[57] For example, environmental liability policies reward policyholders with premium discounts if they take loss prevention measures, such as replacing fuel tanks constructed of corrosion-prone material; by contrast, the premium will be raised by 10 to 20 percent due to a prior leak of the fuel tank.[58] Risk-based pricing, therefore, is a straightforward tool to reduce moral hazard.

In the field of catastrophe insurance, risk-based premiums enable insurers to provide discount to residents who adopt cost-effective mitigation measures, and thus they provide a clear signal to people residing in hazard-prone areas.[59] This also encourages homeowners who plan to settle in hazard-prone areas to reconsider their choice of location and to reduce their vulnerability to catastrophes.[60] Such regulation may not work if insurance premiums are not risk-based. For insurers, furthermore,

---

[54] Omri Ben-Shahar and Kyle D. Logue, *Outsourcing Regulation: How Insurance Reduces Moral Hazard*, 111 Michigan Law Review 197 (2012).

[55] Peter Molk, *Private Versus Public Insurance for Natural Hazards: Individual Behavior's Role in Loss Mitigation* in Risk Analysis of Natural Hazards 265 (Paolo Gardoni et al. eds., 2015).

[56] Feature rating means that insurers examine the insured's individual risk characteristics and adjust premiums accordingly; experience rating means that insurers gather information about the insured's loss experience during the course of the policy period, and use that information either to make retroactive pricing adjustments or prospective pricing adjustments for future policy periods. See Omri Ben-Shahar and Kyle D. Logue, *Outsourcing Regulation: How Insurance Reduces Moral Hazard*, 111 Michigan Law Review 197 (2012).

[57] Tom Baker and Rick Swedloff, *Regulation by Liability Insurance: From Auto to Lawyers Professional Liability*, 60 UCLA Law Review 1412 (2013).

[58] Haitao Yin, Howard Kunreuther and Matthew White, *Risk-Based Pricing and Risk-Reducing Effort: Does the Private Insurance Market Reduce Environmental Accidents?*, 54 Journal of Law and Economics 325 (2011).

[59] Howard Kunreuther, *Oversight of the SBA's Disaster Assistance Program and Examining Changes Proposed by H.R. 3042—The Disaster Loan Fairness Act of 2011*, Testimony before the Subcommittee on Economic Growth, Tax and Capital Access of the House Small Business Committee (2012), available at http://smallbusiness.house.gov/uploadedfiles/kunreuther_testimony.pdf.

[60] Howard Kunreuther and Erwann Michel-Kerjan, *Managing Catastrophic*

risk-based pricing not only assures adequate returns to investors, but also helps to guarantee solvency when catastrophes occur.[61] The relationship with public regulation is clear: to the extent that public regulation would prevent insurers from charging risk-based premiums, this tool aiming at disaster risk reduction could not be employed in an optimal manner.

B.   Contract Design

Contract design can also be used to regulate risk both directly and indi-rectly, by including such elements as deductibles, copayments, coverage amount limit, and exclusions. Deductibles and copayments can mitigate moral hazard directly because they prevent policyholders from shielding themselves entirely from loss.[62] This is one of the tools to control moral hazard: exposing the insured partially to risk will provide incentives for adequate prevention of risks. If a portion of the risk remains with the insured, he will exercise greater vigilance.[63] Exclusions can be seen as an indirect way to regulate policyholders, as they exclude certain types of risk or claim from coverage. Intentional harm, for example, is com-monly excluded from liability insurance policies; environmental claims are also often excluded from general liability insurance (CGL) policies. Deductibles are, moreover, a good technique to remedy adverse selection: policyholders who choose a higher deductible have a lower exposure to risk.

Moreover, using the tools of contract design places a lower burden of information on insurers than when using risk-based pricing since there is no need for actuarial studies to be carried out. It may also be com-paratively efficient in attracting insureds to adopt cheap measures of risk

---

Risks through Redesigned Insurance: Challenges and Opportunities in Handbook of Insurance 517 (G. Dionne ed., 2013).

[61]   Howard Kunreuther and Erwann Michel-Kerjan, *Managing Catastrophic Risks through Redesigned Insurance: Challenges and Opportunities* in Handbook of Insurance 517 (G. Dionne ed., 2013).

[62]   "Deductibles require insureds to pay a fixed amount 'out of pocket' to cover insured losses before the insurance coverage kicks in to cover insured losses thereafter. Copayments typically require insureds to bear some fraction of each covered loss claim filed by an insured". See Omri Ben-Shahar and Kyle D. Logue, *Outsourcing Regulation: How Insurance Reduces Moral Hazard*, 111 Michigan Law Review 197 (2012); Tom Baker and Rick Swedloff, *Regulation by Liability Insurance: From Auto to Lawyers Professional Liability*, 60 UCLA Law Review 1412 (2013).

[63]   Tom Baker and Rick Swedloff, *Regulation by Liability Insurance: From Auto to Lawyers Professional Liability*, 60 UCLA Law Review 1412 (2013).

mitigation.[64] Again, from a regulatory perspective the ability of insurers to incentivize disaster risk reduction via an optimal contract design may be jeopardized as a result of public regulation (e.g. limiting the amount of the deductible).

C. Loss Prevention

Providing loss prevention services is an obvious form of regulation because it permits insurers to advise policyholders on how to modify their behavior in order to mitigate and avoid losses.[65] In other words, loss prevention services can serve as ex ante regulation by insurance. Insurers are better placed than policyholders in identifying the best ways to mitigate risk and avoid losses because they are able to collect more data on claims and harms. Insurers, therefore, will eventually benefit from loss prevention services because they have to pay for the loss based on the policy. Additionally, active engagement in loss prevention will enable insurers to identify insureds with lower than average moral hazard, and to underwrite "good" risks.

Insurers can promote loss prevention in a variety of ways, all of which are potentially applicable to catastrophe insurance.[66] Insurers may monitor the insureds through loss prevention during the course of the insurance relationship;[67] they may conduct research and disseminate new loss-prevention methods;[68] they may cooperate with the State, and promote the legislation of loss prevention laws and regulations;[69] and they

---

[64] Ronen Avraham, *The Law and Economics of Insurance Law—A Primer*, 19 Connecticut Insurance Law Journal 29 (2012).

[65] Tom Baker and Rick Swedloff, *Regulation by Liability Insurance: From Auto to Lawyers Professional Liability*, 60 UCLA Law Review 1412 (2013).

[66] George M. Cohen, *Legal Malpractice Insurance and Loss Prevention: A Comparative Analysis of Economic Institutions*, 4 Connecticut Insurance Law Journal 305 (1997); Tom Baker and Thomas O. Farrish, *Liability Insurance and the Regulation of Firearms* in Suing the Gun Industry: A Battle at the Crossroads of Gun Control and Mass Torts 295 (Timothy D. Lytton ed., 2005).

[67] For example, in the auto insurance context, insurers monitor the insureds' repair service to mitigate loss.

[68] For example, the homeowners' insurance industry has its own association (The Insurance Institute for Business and Home Safety) researching and promulgating various ways of making commercial properties and homes safer from all sorts of hazards.

[69] For example, insurers attempt to upgrade and enhance the content and enforcement of state and local building codes.

may establish underwriting procedures that make loss prevention activities a precondition to obtaining insurance.[70]

Hurricane Andrew changed the manner in which insurers use prevention loss services in catastrophe insurance. Before that catastrophic event, insurers did not promote loss prevention services because they thought these services would prevent them from raising premiums and increasing their profits.[71] Following Hurricane Andrew, the situation changed, and insurers have taken a new approach to loss prevention services: they now feel that they have the potential to initiate fundamental behavioral change among the insureds. The hurricane, moreover, has led insurers to engage with laws and regulations, as it was understood that the severe loss of houses incurred by disasters is due to lack of enforcement of building codes.[72] Consequently, the Insurance Institute for Property Loss Reduction (now the Insurance Institute for Business & Home Safety) was established to promote building codes inspection and enforcement, and to initiate a Code Effectiveness Grading Schedule, which uses the Fire Suppression Rating Program as a prototype.[73] The insurers' approach has also changed from a financial point of view, because they have to demand high premiums in order to underwrite highly risky activities which will lead them to a disadvantaged position in market competition.[74]

Some public-private partnership catastrophe insurance programs expand loss prevention services by providing information to insureds on the benefits of risk mitigation. The legislation for both the Florida Hurricane Catastrophe Fund and the California Earthquake Authority, for example, demands that insurers promote loss prevention services to their clienteles.[75]

In the context of climate change, insurers have worked in tandem with

---

[70]   For example, most insurance policies require that the insureds take all reasonable post-accident activities to mitigate losses or else forfeit coverage. See Robert H. Jerry, II and Douglas R. Richmond, Understanding Insurance Law 637 (4th edn, 2007).

[71]   Robert Hunter, *Insuring Against Natural Disaster*, 12 NAIC Journal of Insurance 467 (1994).

[72]   Howard Kunreuther, *Mitigating Disaster Losses through Insurance*, 12 Journal of Risk and Uncertainty 171 (1996).

[73]   Howard Kunreuther, *Mitigating Disaster Losses through Insurance*, 12 Journal of Risk and Uncertainty 171 (1996).

[74]   Anastasia Telesetsky, *Insurance as a Mitigation Mechanism: Managing International Greenhouse Gas Emissions through Nationwide Mandatory Climate Change Catastrophe Insurance*, 27 Pace Environmental Law Review 691 (2010).

[75]   Dwight Jaffee, *Catastrophe Insurance* in Research Handbook on the Economics of Insurance Law 160 (D. Schwarcz and P. Siegelman eds., 2015).

scientists to identify technical and economic parameters of catastrophe risk, and develop system-wide technologies of loss prevention.[76] In addition, in order to realize the goal of loss prevention, insurers offer low premiums for low emissions operators as an incentive to adopt certain technologies and gradually reduce their emissions.[77]

D. Claim Management

In addition to ex ante regulation such as loss prevention services, insurers also conduct ex post regulation through claim management. Generally speaking, policyholders are usually not concerned about the cost of a claim, leaving its management in the hands of the insurers.[78] Different lines of insurers operate different types of claim management. Liability insurers, due to their right and duty to defend and settle, can directly regulate the litigation process, and thus mitigate ex post moral hazard. Because of their involvement in claim management and litigation, they can further apply such information in pricing, contract design, and loss prevention services. In workers' compensation insurance, since the employer bears the actual risk, insurers are only providing claims administration services based on their expertise in verifying, quantifying, and managing the claims and payments.[79]

In the case of catastrophe risks, the policyholders' inability to change the possibility of a natural disaster, alongside their ability to mitigate disaster losses, makes claim management necessary to control ex post moral hazard. Catastrophe insurers, therefore, may employ an adjuster to investigate claimed losses, measure them, and negotiate payouts. They can then review the adjuster's decisions, and provide greater uniformity and predictability.

E. Refusal to Insure

This is the last, but not least, technical tool used by insurers to regulate the insured, and it is especially effective when some activities cannot

---

[76] P.A. Scott et al., *Human Contribution to the European Heat Wave of 2003*, 432 Nature 610 (2004).

[77] Anastasia Telesetsky, *Insurance as a Mitigation Mechanism: Managing International Greenhouse Gas Emissions through Nationwide Mandatory Climate Change Catastrophe Insurance*, 27 Pace Environmental Law Review 691 (2010).

[78] Tom Baker, *Liability Insurance Conflicts and Defense Lawyers: From Triangles to Tetrahedrons*, 4 Connecticut Insurance Law Journal 101, 107 (1997).

[79] Omri Ben-Shahar and Kyle D. Logue, *Outsourcing Regulation: How Insurance Reduces Moral Hazard*, 111 Michigan Law Review 197, 213–214 (2012).

be undertaken without insurance. There may be an obligation to take out insurance based either on regulation (mandatory insurance) or on contract. An example of the latter is homeowners' insurance in the United States. Before a mortgage contract is concluded, the mortgagor is required to obtain homeowners' insurance or to relegate to the mortgagee to acquire such insurance.[80] In this case, and presuming that catastrophe insurance is mandatory in high hazard-prone areas, the insurers' refusal to insure may have de facto control over the insureds, and may induce less risky behavior. A refusal to insure is more complex in some legal systems, such as in France,[81] where the purchase of disaster cover is mandatory, or is at least a mandatory complement to a voluntary homeowners insurance. In that case insurers are often forced by regulation to provide cover and refusal to insure is then no longer an option.

Declining to renew a policy is another form of refusal to insure, and can be equally effective. After the insured has conducted risky activities or failed to take mitigation measures, the insurers can cancel, rescind, or refuse to renew the existing policy.[82] The threat of non-renewal could push homeowners to undertake mitigations.

IV.   REGULATION BY CATASTROPHE INSURANCE: EXAMPLES

The regulatory techniques of the insurance industry identified in the previous section have already been put into effect in different countries. This section examines regulation by the private flood insurance system in the United Kingdom, one of only a handful of successful examples in the world; regulation by the National Flood Insurance Program (NFIP) in the United States; regulation by the Catastrophes Naturelles (Cat.Nat.) insurance system in France, which is mandatory with property insurance; regulation by the Japanese Earthquake Reinsurance Scheme (JER), which is voluntary for policyholders but mandatory for insurers; and regulation by the Turkish Catastrophe Insurance Pool (TCIP), which is "considered as a good example of catastrophe risk insurance for developing and

---

[80]   Howard Kunreuther, *Has the Time Come for Comprehensive Natural Disaster Insurance?* in On Risk and Disaster: Lessons from Hurricane Katrina 175 (Ronald J. Daniels et al. eds., 2006).

[81]   See section IV.C below.

[82]   Omri Ben-Shahar and Kyle D. Logue, *Outsourcing Regulation: How Insurance Reduces Moral Hazard*, 111 Michigan Law Review 197, 213–214 (2012).

middle-income countries".[83] As these examples will demonstrate, there is wide variation in the nature and extent of regulation through catastrophe insurance across different countries. For each country I will first generally sketch the availability of catastrophe insurance and whether this is influenced by public regulation; next, I will examine to what extent the technical tools I discussed in the previous section (risk-based pricing, contract design, loss prevention, claim management, or refusal to insure) can and are used in practice.

A.   United Kingdom

In the United Kingdom, natural catastrophe risks coverage is included among the basic guarantees in commercial and household policies. Many households, for example, are in effect covered against flood damage, which is usually included in homeowners' insurance policies, because mortgage lenders require that a property have full insurance coverage.[84] The flood insurance scheme emerged in 1961. According to a gentlemen's agreement[85] that divided the rights and duties between the State and the insurance industry, insurers regulate policyholders and compensate victims in the case of flood damage, while the State sets rules and codes for flood protection, flood warning, and land use, and guarantees the independence of insurers.[86] The distinguishing feature of the UK's catastrophe insurance scheme was that the State did not intervene in either direct insurance or reinsurance. This UK model was based on a close collaboration between the State and private insurers whereby the private insurers agreed to provide insurance cover and the State committed to invest in flood protection prevention measures. For a long time this UK private flood insurance scheme was considered as a model showing how a largely private insurance industry could work in an efficient and sustainable

---

[83]   J. Bommer et al., *Development of an Earthquake Loss Model for Turkish Catastrophe Insurance*, 6(3) Journal of Seismology 431 (2002).

[84]   Erwann Michel-Kerjan, *Catastrophe Economics: The National Flood Insurance Program*, 24(4) Journal of Economic Perspectives 165 (2010).

[85]   According to a gentlemen's agreement between the British government and private insurers, the insurers undertook to offer flood coverage to owners of houses and organizations. See Michael Huber, *Reforming the UK Flood Insurance Regime: The Breakdown of a Gentlemen's Agreement*, CARR Discussion Papers, DP 18 (2014).

[86]   Michael Huber, *Insurability and Regulatory Reform: Is the English Flood Insurance Regime Able to Adapt to Climate Change?*, 29 The Geneva Papers on Risk and Insurance—Issues and Practice 169 (2004).

manner.[87] However, recent floods have fundamentally challenged the mechanisms because it has been claimed (by insurers and public opinion) that the State did not sufficiently invest in flood protection measures and therefore was not adhering to its part of the deal. Insurers even threatened cancelling the gentlemen's agreement. These recent evolutions therefore fundamentally challenged the stability of the system.[88]

In 2013, the State and the insurance industry agreed to a Memorandum of Understanding (MOU) known as Flood Re,[89] which has its basis in the UK Government Water Act 2014. Flood Re, a not-for-profit reinsurance fund owned and managed by private insurers, is designed to ensure regulation by flood insurance and keep it widely available and affordable, specifically to enable high-flood-risk households to obtain insurance at an affordable price. It is estimated that between 300,000 and 500,000 high-flood-risk households would struggle to obtain affordable flood insurance without Flood Re.[90] The Water Act 2014 lays the foundations for the detailed provisions as to the structure and working of the scheme, which was launched on 4 April 2016. Primary insurers sell a homeowners' insurance policy, which contains flood coverage, to households in the usual way, and then pass the flood risk to Flood Re, which pays the insurers if flood claims are made.[91] The scheme ensures regulation by flood insurance because the claim still rests with the primary insurers, but they are backed up by Flood Re. The Flood Re fund has two sources of income: one is the flood element premium of the home insurance policies,

---

[87] Michael Huber, *Insurability and Regulatory Reform: Is the English Flood Insurance Regime Able to Adapt to Climate Change?*, 29 The Geneva Papers on Risk and Insurance—Issues and Practice 169 (2004).

[88] Johanna Hjalmarsson and Mateusz Bek, *Legislative and Regulatory Methodology and Approach: Developing Catastrophe Insurance in China* in Insurance Law in China 196 (Johanna Hjalmarsson and Dingjing Huang eds., 2015).

[89] The Flood Re model is loosely based on Pool Re, a reinsurance scheme for terrorism risks formed in 1993 in response to the threat posed by the Irish Republican Army and other terrorist activity. See Johanna Hjalmarsson and Mateusz Bek, *Legislative and Regulatory Methodology and Approach: Developing Catastrophe Insurance in China* in Insurance Law in China 197 (Johanna Hjalmarsson and Dingjing Huang eds., 2015).

[90] Association of British Insurers (ABI), *The Future of Flood Insurance: What Happens Next?* (2015) available at https://www.abi.org.uk/Insurance-and-savings/Topics-and-issues/Flooding/Government-and-insurance-industry-flood-agreement/The-future-of-flood-insurance.

[91] Association of British Insurers (ABI), *Flood Re Explained* (2015) available at https://www.abi.org.uk/Insurance-and-savings/Topics-and-issues/Flooding/Government-and-insurance-industry-flood-agreement/Flood-Re-explained.

and the other is an additional levy on the insurance industry. However, in an extreme situation—for example, a year with damages six times worse than 2007—the government will take primary responsibility, and work with both the insurers and Flood Re.[92]

Risk-based pricing Initially, premiums were undifferentiated across all households, but insurers gradually improved their knowledge through accurate flood maps, and took the real risks into account. This is important since premiums of flood insurance are risk-based, not flat, and are set on a case-by-case basis.[93] In 2001, for example, heavy premiums were loaded for properties where flood claims have been previously made.[94] Furthermore, for households located in high flood-prone areas, premiums have increased significantly during the last few years.[95]

Insurers in the United Kingdom prefer to conduct risk-based pricing of flood insurance because, first, the State lacks control over the rate-setting as per the gentlemen's agreement;[96] second, it helps control moral hazard, and "bad" risks are sorted out more rigorously; third, it may provide incentives to policyholders to mitigate flood risks. Flood Re has also been criticized because high-risk houses will de facto be subsidized through a levy that will have to be paid by all domestic property owners.[97]

Contract design Deductibles are applied to all or some indemnification,

---

[92] Association of British Insurers (ABI), Government and Insurance Industry Flood Agreement (Statement of Principles) (2014) available at https://www.abi.org.uk/Insurance-and-savings/Topics-and-issues/Flooding/Government-and-insurance-industry-flood-agreement.

[93] Michael Huber and Amodu Tola, *United Kingdom* in Financial Compensation for Victims of Catastrophe: A Comparative Legal Approach 291 (M. Faure and T. Hartlief eds., 2006).

[94] David Crichton, *UK and Global Insurance Responses to Flood Hazard*, 27 Water International 119 (2002).

[95] J.E. Lamond et al., *Accessibility of Flood Risk Insurance in the U.K.: Confusion, Competition and Complacency*, 12(6) Journal of Risk Research 825 (2009).

[96] But the insurers also "agreed that the additional premium rate would not exceed 0.5 percent on the sum insured". See David Crichton, *UK and Global Insurance Responses to Flood Hazard*, 27 Water International 119 (2002). What is more, according to the agreement between insurers and the government to develop the non-profit company Flood Re, insurers will charge high-risk households a premium that will be capped depending on the property's Council Tax band. See Association of British Insurers (ABI), Flood Re Explained (2015) available at https://www.abi.org.uk/Insurance-and-savings/Topics-and-issues/Flooding/Government-and-insurance-industry-flood-agreement/Flood-Re-explained.

[97] J. Davy, *Flood Re Risk Classification and Distortion of the Market* in Future Directions of Consumer Flood Insurance in the UK. Reflections upon the Creation of Flood Re 26 (J. Hjalmarsson ed., 2015).

depending on the type of damage and its cause. This follows the model provided by building insurance and content insurance, which cover not just ordinary perils like fire, but also earthquakes, floods and other catastrophe risks.[98] Individual policy deductibles per 10^5 IV is 1 percent (between 78 and 156 on average, but could reach up to 2,333).[99] Exclusions are also utilized in Flood Re, as homes built after January 1, 2009 will not be covered if they are constructed in known high-flood-risk areas (as applied under the old Flood Insurance Statement of Principles).[100] Such an arrangement offers real-estate developers the incentives to avoid construction in known high-flood-risk areas.

Loss prevention Insurers promote loss prevention in a variety of ways. First, insurers actively engage with government regulation. In 2007, the Association of British Insurers (ABI) demanded more government involvement in flood-risk reduction, the approval of new compulsory building codes, and the development of long-term (25 years) preventive strategy plans.[101] Recently, the State has created Planning Policy Statement (PPS) 25 in collaboration with insurers, which proscribes land-use planning and flood damage reduction.[102] Second, insurers conduct catastrophe risk research. At least 12 major insurers have invested substantial sums in research aimed at producing more accurate flood maps. Although such research is expensive, these maps, which are better than the UK government or its agencies have been able to afford so far, will assist insurers to underwrite, and lead to more accurate pricing.[103]

---

[98]   World Forum of Catastrophe Programmes, *Natural Catastrophes Insurance Cover: A Diversity of Systems* 176 (2008) available at http://.wfcatprogrammes. com/c/document_library/get_file?folderId=13442&name=DLFE-553.pdf.

[99]   The amounts for individual policy deductibles per 10^5 IV and premium levels are assessed on the basis of maximum damage (i.e., in case a house is completely destroyed). See Youbaraj Paudel, *A Comparative Study of Public–Private Catastrophe Insurance Systems: Lessons from Current Practices*, 37 The Geneva Papers on Risk and Insurance—Issues and Practice 257 (2012).

[100]   Association of British Insurers (ABI), *The Future of Flood Insurance: What Happens Next?* (2015) available at https://www.abi.org.uk/Insurance-and-savings/Topics-and-issues/Flooding/Government-and-insurance-industry-flood-agreement/The-future-of-flood-insurance.

[101]   Association of British Insurers (ABI), *Summer Floods 2007: Learning the Lessons* (2015), available at http://www.ambiental.co.uk/downloads/ABI_2007_Summer_Floods_Review.pdf.

[102]   Michael Huber, *Insurability and Regulatory Reform: Is the English Flood Insurance Regime Able to Adapt to Climate Change?*, 29 The Geneva Papers on Risk and Insurance—Issues and Practice 169 (2004).

[103]   David Crichton, *UK and Global Insurance Responses to Flood Hazard*, 27 Water International 119 (2002).

Claim management Under the private insurance scheme, claims are made via the insurance company, and are established in individual insurance contracts. Because data gathering is focused on claim histories, and experience rating is applied in risk-based pricing, claim management helps control moral hazard.

Refusal to insure Individuals and organizations have a de facto obligation to buy flood coverage if they want to secure a mortgage credit, because all homeowners wishing to secure a mortgage credit must purchase flood insurance.[104] If the properties lack insurance coverage, they may decrease in value even when they are no longer marketable.[105] Such quasi-mandatory arrangement makes the insureds take more than normal precautions. Therefore, insurers' refusal to insure will all but control the insureds' activities, and insurers can use this power to induce less risky behavior. As mentioned above, a consequence of the gentlemen's agreement is that private insurers in principle undertake to offer flood coverage to owners of houses and organizations.[106] That, however, does not imply an unconditional commitment to provide cover for any risk.

Furthermore, insurers may refuse to renew flood policies, and negotiate with the government to undertake stronger protection measures. Indeed, the ABI once warned the government to take firmer action on flood defense, otherwise the insurance industry would not be able to provide flood coverage.[107] This has become clear in recent years where many floods occurred and insurers claimed that this was due to the lack of investments by the government in flood prevention, as a result of which insurers wanted to cancel the gentlemen's agreement. If insurers are entitled to withdraw from the market, the problems of catastrophe risk will eventually be left for the State and society to resolve.

B.   United States

The United States is often seen as an insurance-based society, whereby there are strong interdependencies between the government and the

---

[104]   Michael Huber, *Reforming the UK Flood Insurance Regime: The Breakdown of a Gentlemen's Agreement*, CARR Discussion Papers, DP 18 (2014).

[105]   Michael Huber, *Insurability and Regulatory Reform: Is the English Flood Insurance Regime Able to Adapt to Climate Change?*, 29 The Geneva Papers on Risk and Insurance—Issues and Practice 169 (2004).

[106]   Michael Huber, *Reforming the UK Flood Insurance Regime: The Breakdown of a Gentlemen's Agreement*, CARR Discussion Papers, DP 18 (2014).

[107]   David Crichton, *UK and Global Insurance Responses to Flood Hazard*, 27 Water International 119 (2002).

insurance industry.[108] This government involvement can also be observed in relation to catastrophic risks. Three distinct models of collaboration between the government and the insurance sector can be distinguished.[109] In the first model, private insurers are the principal guarantors against risk, and the government has only limited involvement. The Price-Anderson Act, concerning nuclear facilities, is an example of this model.[110] The Price-Anderson Act only mandated the purchase of insurance but (since 1975) there is no longer government involvement in the compensation element. A first layer is paid by the operator's liability insurer where the accident occurred; the second layer is provided through a collective payment by all nuclear operators active in the market through retroactive premiums collected by the Nuclear Regulatory Commission (NRC). The NRC manages the collection of the retrospective premiums, but the financial risk is born by the nuclear operators.[111] In the second model, insurers provide the primary coverage for the risk while the State supplies the reinsurance coverage. The Federal Terrorism Risk Program illustrates this model.[112] In the third model, insurers do not assume risks, but only administer policy coverage for government agencies. Earthquake insurance in California (California Earthquake Agency)[113] and the National

---

[108]   Tom Baker and Thomas O. Farrish, *Liability Insurance and the Regulation of Firearms* in Suing the Gun Industry: A Battle at the Crossroads of Gun Control and Mass Torts 292 (Timothy D. Lytton ed., 2005).

[109]   Robert L. Rabin and Suzanne A. Bratis, *United States* in Financial Compensation for Victims of Catastrophe: A Comparative Legal Approach 324 (M. Faure and T. Hartlief eds., 2006).

[110]   42 U.S.C. 2210 (1988 & Supp. 1992). "Today, the individual liability of a nuclear operator is $375 million supplemented with a second layer of retrospective premiums of $11.86 billion, leading to a total amount of $12.2 billion without any government intervention". See Liu Jing and Michael Faure, *Compensating Nuclear Damage in China*, 11 Washington University Global Studies Law Review 781, 813 (2012).

[111]   Michael Faure and Tom Vanden Borre, *Compensating Nuclear Damage: A Comparative Economic Analysis of the U.S. and International Liability Schemes*, 33 Williamm & Mary Environmental Law and Policy Review 219, 240–247 (2008).

[112]   Veronique Bruggeman, Michael Faure and Tobias Heldt, *Insurance Against Catastrophe: Government Stimulation of Insurance Markets for Catastrophic Events*, 23 Duke Environmental Law and Policy Forum 185, 230–231 (2012).

[113]   The California Earthquake Authority (CEA) is a state-run privately funded earthquake insurance program. Earthquake insurance can be purchased for an additional premium in all states except California, where it is usually necessary to buy an earthquake policy for residential damage through the CEA. See Veronique Bruggeman, Michael Faure and Tobias Heldt, *Insurance Against Catastrophe: Government Stimulation of Insurance Markets for Catastrophic Events*, 23 Duke Environmental Law and Policy Forum 185, 224–225 (2012).

Flood Insurance Program (NFIP) follow this model.[114] The following discussion will focus on the third model of natural catastrophe risks.

The United States is vulnerable to several natural catastrophes, and the risk of loss is increasing significantly.[115] Standard homeowners and commercial insurance policies normally cover non-catastrophe damage, such as fire, wind, hail, and lightning; however, flood damage resulting from rising water and earthquakes (in California) is normally explicitly excluded from coverage.[116] Flood insurance was first offered by private insurers in the late 1890s, yet the financial loss was too large for insurers, who consequently left the market.[117] The NFIP, administered by the Federal Emergency Management Agency (FEMA), was established according to the National Flood Insurance Act of 1968, in order to assume the flood risk and offer coverage.[118] The Standard Flood Insurance Policy of the NFIP covers direct physical losses to structures and their contents caused by flood.[119] The NFIP has sold more than 5.2 million policies in 22,000 communities in the past 40 years, and provided almost $1.3 trillion in coverage.[120] Most of these policies are for single-family, residential properties—such as found in Florida—which comprise nearly 40 percent

---

[114] Véronique Bruggeman, Compensating Catastrophe Victims: A Comparative Law and Economics Approach 415–432 (2010).

[115] US Government Accountability Office, *Catastrophe Insurance Risk: The Role of Risk-linked Securities and Factors Affecting Their Use* 8 GAO-02-941 (2002). According to the Federal Emergency Management Agency (FEMA), an event where related federal costs reach or exceed $500 million is deemed as "catastrophe". See US Government Accountability Office, *Experiences from Past Disasters Offer Insights for Effective Collaboration after Catastrophe Events* 2 GAO-09-811 (2009). See also Michel-Kerjan, Erwann, Jeffrey Czajkowski and Howard Kunreuther, *Could Flood Insurance be Privatized in the United States? A Primer*, 40 The Geneva Papers on Risk and Insurance Issues and Practice 179 (2015).

[116] Seema Patel and Sarala Nagala, *Public Policy Considerations of Water Damage Exclusions in Hurricane Insurance Policies* 17, available at https://www.law.berkeley.edu/library/resources/disasters/Patel_Nagala.pdf.

[117] Howard Kunreuther and Richard J. Roth, Sr, Paying the Price: The Status and Role of Insurance Against Natural Disasters in the United States 40 (1998).

[118] But some private insurers still offer excess flood protection that provides higher limits of coverage than the NFIP. See B. Well, *Excess Flood Market Steps Up When National Flood Program Falls Short*, Insurance Journal (July 24, 2006).

[119] Robert L. Rabin and Suzanne A. Bratis, *United States* in Financial Compensation for Victims of Catastrophe: A Comparative Legal Approach 332 (M. Faure and T. Hartlief eds., 2006).

[120] Howard Kunreuther, *The Role of Insurance in Reducing Losses from Extreme Events: The Need for Public–Private Partnerships*, 40 The Geneva Papers on Risk and Insurance—Issues and Practice 741 (2015).

of the NFIP (in number of policies, premiums and coverage).[121] However, due to homeowners' underestimation of the likelihood of flood damages, the penetration rate of flood insurance is not very high. For example, only 20 percent of those who suffered damage from Hurricane Sandy had purchased NFIP policies.[122]

FEMA, in administrating the NFIP, works in conjunction with private insurance companies through the "Write Your Own" (WYO) program, which allows private insurers to issue policies in their own name, to adjust flood claims, and to defend, settle or pay all claims arising from the flood policies.[123] Moreover, there is no reinsurance arrangement in the NFIP, and if claims exceed its financial capacity, the federal government provides bailout. For example, after Hurricane Katrina, the NFIP required a bailout from the US Treasury of close to $20 billion.[124] Through these cooperative efforts by the insurance industry and the government—where private insurers make use of their marketing channels, risk management expertise, and existing policy base, and the federal government works as the ultimate risk taker—the NFIP enables homeowners to purchase available flood insurance.

Risk-based pricing Premium setting in the NFIP is partially risk-based. At the very beginning, the NFIP tried to adopt risk-based premiums that differ per flood zone, but this proved to be difficult in practice. Because the owners of buildings built before the creation of the NFIP are reluctant to purchase higher policies, premiums are determined by applying the Actuarial Rate Formula. The NFIP's overall pricing strategy, however, leads to important divergences from the true risk for a number of residents covered by the program.[125] In 2012, the Biggert-Waters Flood Insurance Reform Act eliminated certain premium subsidies and increased risk-based pricing. However, in 2014, this was prohibited by the Homeowner

---

[121] Erwann Michel-Kerjan and C. Kousky, *Come Rain or Shine: Evidence on Flood Insurance Purchases in Florid*, 77 Journal of Risk and Insurance 369 (2010).
[122] Christopher C. French, *Insuring Floods: The Most Common and Devastating Natural Catastrophes in America*, 60 Villanova Law Review 53 (2015).
[123] Véronique Bruggeman, Compensating Catastrophe Victims: A Comparative Law and Economics Approach 420 (2010); Robert L. Rabin and Suzanne A. Bratis, *United States* in Financial Compensation for Victims of Catastrophe: A Comparative Legal Approach 331 (M. Faure and T. Hartlief eds., 2006).
[124] Dwight Jaffee, *Catastrophe Insurance*, in Research Handbook on the Economics of Insurance Law, 160 (D. Schwarcz and P. Siegelman eds., 2015).
[125] Erwann Michel-Kerjan, Jeffrey Czajkowski and Howard Kunreuther, *Could Flood Insurance be Privatized in the United States? A Primer*, 40 The Geneva Papers on Risk and Insurance—Issues and Practice 179 (2015).

Flood Insurance Affordability Act, which restored grandfathering and limited certain rate increases.

According to the calculation of Michel-Kerjan et al., around a quarter of the total NFIP policies today are subsidized.[126] Subsidized premiums do not reflect the accurate flood risk, and represent on average only 35–50 percent of the actual risk.[127] Moreover, subsidized structures are generally more prone to flooding, and are thus riskier than other risk-based premiums structures.[128]

Contract design The NFIP provides deductibles, ranging from between $500 and $5,000. Although a higher deductible lowers the premium and encourages more mitigation measures, "97 percent of NFIP policy-holders choose deductible levels of $1000 or less".[129] The NFIP also uses coverage limits. For example, a single family dwelling is normally eligible for up to $250,000 in building coverage and up to $100,000 in personal property coverage.[130]

Loss prevention The National Flood Insurance Reform Act of 1994 creates mitigation insurance and develops a mitigation assistance program for the NFIP. The NFIP integrates risk mitigation and prevention measures, and administers different kinds of mitigation programs. For example, the NFIP tries to supply premium discount to encourage mitigation of risk. It operates the Community Rating System (CRS), which rewards communities that undertake mitigating activities with premiums discounts.[131]

Although the NFIP successfully reduced the vulnerability of new buildings to floods, it had less impact on existing buildings and was not able to limit the development of flood-prone areas.[132] The increasing federal

---

[126]   Erwann Michel-Kerjan, Jeffrey Czajkowski and Howard Kunreuther, *Could Flood Insurance be Privatized in the United States? A Primer*, 40 The Geneva Papers on Risk and Insurance—Issues and Practice 179 (2015).

[127]   US Government Accountability Office, *Federal Emergency Management Agency: On-going Challenges Facing the National Flood Insurance Program*, GAO-08-118T (2007).

[128]   US Government Accountability Office, *Flood Insurance: Strategies for Increasing Private Sector Involvement*, GAO-14-127 (2014).

[129]   Erwann Michel-Kerjan and C. Kousky, *Come Rain or Shine: Evidence on Flood Insurance Purchases in Florid*, 77 Journal of Risk and Insurance 369 (2010).

[130]   Robert L. Rabin and Suzanne A. Bratis, *United States* in Financial Compensation for Victims of Catastrophe: A Comparative Legal Approach 332 (M. Faure and T. Hartlief eds., 2006).

[131]   Paul Hudson et al., *Moral Hazard in Natural Disaster Insurance Markets: Empirical Evidence from Germany and the United States*, 93 Land Economics 179 (2017).

[132]   R.J. Burby, *Rising Tide: The Great Mississippi Flood of 1927 and How It Changed America*, 66(3) Journal of the American Planning Association 337 (2000).

disaster relief, moreover, may reduce individuals' incentive to prevent loss and may contribute to this result.[133] There has been substantial criticism of the payments made after Hurricane Katrina, arguing that they will encourage people to rebuild in vulnerable areas.[134] Some hold that the NFIP therefore provides incentives for property development in high-risk areas.[135]

Claim management The NFIP uses insurers, because of their claims handling expertise, to settle claims on its behalf. Yet the NFIP bears further responsibility with regards to claim management, as the Flood Insurance Reform Act of 2004 stipulates that it should increase and improve guidance for policyholders about the flood insurance claims process, and reduce compensation to properties for which repetitive flood insurance claim payments have been made. However, anecdotal evidence suggests that because insurers do not assume underwriting risk in the NFIP, the claims costs are higher than they would be under a private insurance scheme.[136]

Refusal to insure This regulatory tool has little function in the NFIP. Since insurers do not assume underwriting risk and receive an expense allowance for policies written, they have no incentives to refuse to insure. Instead, the NFIP makes every effort to attract individuals to subscribe to the flood insurance policy. The Flood Disaster Protection Act of 1973 mandates that lenders require flood insurance on loans secured by properties that are located within high-risk flood areas.[137] Moreover, the National Flood Insurance Reform Act of 1994 prevents federal agencies

---

[133]   The number of Presidential disaster declarations has significantly increased over the past 50 years: namely, from 162 over the period 1955–1965 to 545 during 1996–2005. In response to Hurricane Katrina in 2005 and in the following year, three emergency supplemental appropriations bills of approximately $88.4 billion were enacted by Congress. This total amount of federal relief is more than the combined total amounts of private wind insurance claims and NFIP claims. See E. Michel-Kerjan, S. Lemoyne de Forges and H. Kunreuther, *Policy Tenure under the US National Flood Insurance Program (NFIP) USA*, 32 Risk Analysis 644 (2012).

[134]   W.F. Shughart, *Katrinanomics: The Politics and Economics of Disaster Relief*, 127 Public Choice 31, 44 (2006).

[135]   A.T. Young, *Replacing Incomplete Markets with a Complete Mess: Katrina and the NFIP*, 35 International Journal of Social Economics 561, 566 (2008); Justin Pidot, *Deconstructing Disaster*, BYU Law Review 213 (2013).

[136]   Because when the payment of claims exceeds their premium funds, they can collect FEMA letters of credit for any claim amount. See David Crichton, *UK and Global Insurance Responses to Flood Hazard*, 27 Water International 119 (2002).

[137]   Carolyn Kousky and Erwann Michel-Kerjan, *Examining Flood Insurance Claims in the United States: Six Key Findings*, 84 Journal of Risk and Insurance 819 (2015).

from granting disaster aid in the Special Flood Hazard Areas (SFHAs) to communities that have not joined the NFIP.[138]

C. France

In France, catastrophic risks, such as floods and earthquakes, were traditionally excluded from insurance coverage. However, after the 1981 floods in the Rhone, Saone and Garonne valleys, the French legislator created the Act of July 13, 1982, which establishes the Catastrophes Naturelles (Cat. Nat) System.[139] This system offers a unique public–private partnership in regulating catastrophic risks. The division of responsibilities between the insurers and the State according to the Cat.Nat System is as follows: insurers are responsible for underwriting policies, managing additional premiums, adjusting damages, handling claims, and paying indemnifications, while the State is responsible for reinsurance and cooperating with insurers to create prevention and mitigation plans.[140] Article 1 of the Act of July 13, 1982 provides that property insurance policies that cover damage against property are automatically and mandatorily insured against the risk of natural disasters. Although natural catastrophe disasters are "non-insurable direct material damage", they must be insured in the Cat.Nat System (Article L. 125-1 Insurance Code).[141] This mandatory requirement, coupled with its efficient enforcement by the French authorities, brings the penetration rate of catastrophe insurance to nearly 100 percent.[142] In addition, the State will back private insurers via reinsurance by the Caisse Centrale De Reassurance (CCR) with unlimited State guarantee.[143] This

---

[138]   World Forum of Catastrophe Programmes, *Natural Catastrophes Insurance Cover: A Diversity of Systems* 185 (2008), available at http://www.wfcatprogrammes.com/c/document_library/get_file?folderId=13442&name=DLFE-553.pdf.
[139]   Act No. 82-600 of July 13, 1982 on the Indemnification of Victims of Natural Catastrophes, JORF, July 14, 1982, 2242.
[140]   Youbaraj Paudel, *A Comparative Study of Public–Private Catastrophe Insurance Systems: Lessons from Current Practices*, 37 The Geneva Papers on Risk and Insurance—Issues and Practice 257 (2012).
[141]   Article L. 125-1: "Non insurable direct material damage the determining cause of which was the abnormal intensity of a natural agent, when normal measures taken to avoid such damage have been unable to prevent the occurrence thereof or could not be taken, shall be deemed to be a natural disaster within the meaning of this chapter".
[142]   Erwann Michel-Kerjan, *Catastrophe Economics: The National Flood Insurance Program*, 24 Journal of Economic Perspectives 165 (2010).
[143]   World Forum of Catastrophe Programmes, *Natural Catastrophes Insurance Cover: A Diversity of Systems* 61 (2008) available at http://www.wfcatprogrammes. com/c/document_library/get_file?folderId=13442&name=DLFE-553.pdf.

enables primary insurers to underwrite catastrophe insurance policies at affordable prices for homeowners.

Risk-based pricing The Cat.Nat System adopts a flat rate rather than risk-based premiums. The government fixes the premiums corresponding to the guarantee against the effects of natural catastrophes. Under the influence of the national solidarity principle, Article 2 of the Act of July 13, 1982 stipulates that "this guarantee is financed by an additional premium calculated on the basis of a single rate set by Decree for each category of insurance policy". This additional premium for catastrophe coverage is decided by the State in the form of a Ministerial Order, and applied to each type of basic policies.[144] Originally the initial rate was 5.5 percent in 1982; this increased to 9 percent the following year and to 12 percent in 2000.[145] As this flat premium does not comply with the principle of risk-based pricing, in principle it creates few incentives for policyholders to reduce risk. However, in theory insurers could still compete as far as the basic premium for housing insurance is concerned, to which the Cat.Nat complement is linked.[146] It is not so clear to what extent this really is the case; moreover, even if there were such a competition it is unclear whether there would be a reward for lower risks and hence a risk-differentiation. In 2006, the French public authorities presented a draft amendment to the 1982 Act, in an attempt to abandon the unique extra insurance premium rate.[147]

Contract design There are mandatory and non-index-linked deductibles fixed in the Act. Originally, the amount of deductibles differed based on the type of risk—residential or commercial—but remained the same for all perils (except subsidence, which has a higher specific deductible). The Decree of August 10, 1982, the Decrees of September 7 and 19, 1983, and the Decree of September 5, 2000 all insist on this rule. However, in order to control moral hazard and encourage loss prevention measures, a sliding scale has been introduced to vary these deductibles since 1 January

---

[144] World Forum of Catastrophe Programmes, *Natural Catastrophes Insurance Cover: A Diversity of Systems* 64 (2008), available at http://www.wfcatprogrammes. com/c/document_library/get_file?folderId=13442&name=DLFE-553.pdf.

[145] Michel Cannarsa, Fabien Lafay and Olivier Moréteau, *France* in Financial Compensation for Victims of Catastrophe: A Comparative Legal Approach 101 (M. Faure and T. Hartlief eds., 2006).

[146] Roger Van den Bergh and Michael Faure, *Compulsory Insurance of Loss to Property Caused by Natural Disasters: Competition or Solidarity?*, 29 World Competition 25 (2006).

[147] World Forum of Catastrophe Programmes, *Natural Catastrophes Insurance Cover: A Diversity of Systems* 64 (2008), available at http://www.wfcatprogrammes. com/c/document_library/get_file?folderId=13442&name=DLFE-553.pdf.

2001.[148] Exclusions are also used in the Cat.Nat System, as the Act of July 13, 1982 stipulates that damage or costs indirectly due to the disaster event are not covered.[149]

Loss prevention The Cat.Nat System integrates risk mitigation and prevention measures. Insurers, moreover, cooperate with the State to formulate risk prevention plans and form the Barnier mitigation fund.[150] The amount of the deductible also depends on whether a particular municipality has adopted a "prevention of risk plan" (*plan de prevention des risques*). This should incentivize the local population to press the municipality to adopt a prevention plan.[151] However, recent empirical evidence shows that this system does not provide optimal incentives for flood damage reduction. The deductibles' adjustment policy does not seem to provide incentives to communities to adopt a risk prevention plan in practice.[152]

Claim management The Insurance Code specifies the legal procedure of claim management. After government authorities declare a "natural catastrophe" in the official gazette, the insureds must report their damage to the insurers within 10 days, together with all relevant documentation including a statement of all direct damage to property (indirect damages

---

[148]   A sliding scale deductible means that if a state of natural catastrophe was declared in the area three times in the previous five years for the same sort of risk (such as floods), deductibles are doubled; if four times, deductibles are trebled, and from five times on, deductibles are multiplied by four. This deductible increase happens when the loss occurs in municipalities without a Foreseeable Natural Risks Prevention Plan. See Véronique Bruggeman, Compensating Catastrophe Victims: A Comparative Law and Economics Approach 307 (2010); World Forum of Catastrophe Programmes, *Natural Catastrophes Insurance Cover: A Diversity of Systems* 65 (2008), available at http://www.wfcatprogrammes.com/c/document_library/get_file?folderId=13442&name=DLFE-553.pdf.

[149]   World Forum of Catastrophe Programmes, *Natural Catastrophes Insurance Cover: A Diversity of Systems* 64 (2008), available at http://www.wfcatprogrammes.com/c/document_library/get_file?folderId=13442&name=DLFE-553.pdf.

[150]   Youbaraj Paudel, *A Comparative Study of Public–Private Catastrophe Insurance Systems: Lessons from Current Practices*, 37 The Geneva Papers on Risk and Insurance—Issues and Practice 257 (2012).

[151]   O. Moréteau, *Policing the Compensation of Victims of Catastrophes: Combining Solidarity, Self-Responsibility* in Shifts in Compensation between Private and Public Systems 199, 217 (W.H. van Boom and M. Faure eds., 2007).

[152]   "In terms of financing damage mitigation measures, since 2005, the Fund for the Prevention of Major Natural Risks, also called the 'Barnier' fund, has been providing subsidies up to €125 million per year for studies on assessments of natural disaster risk and potential prevention and protection measures for buildings". See Jennifer K. Poussin, W.J. Wouter Botzen and Jeroen C.J.H. Aerts, *Stimulating Flood Damage Mitigation through Insurance: An Assessment of the French CatNat System*, 12 Environmental Hazards 258 (2013).

are excluded), photos, videos, etc. (Article L. 125 of the Insurance Code). The timeframe for claim reporting is very strict (except when suspended by force majeure), and non-compliance may exclude the right to compensation.[153] Setting a strict timeframe will press the policyholders to act with due care and diligence after the catastrophe, and allow insurers to send adjusters as soon as possible.

Refusal to insure Although premiums are not risk-based, insurers may not refuse to underwrite individuals' catastrophe risk. When insurers undertake the higher risk, they can reduce risk by purchasing relatively cheap reinsurance policies from the CCR, which is the only reinsurer with unlimited State guarantee.[154]

D. Japan

The current Japanese earthquake insurance system is a public–private partnership between the government and the insurance industry. The system is divided into two different regimes, one for business and industry and the other for households.[155] Business and industrial risks are covered primarily by the private insurance market, while household risks are covered by private insurers, but with strong government involvement.[156]

The household earthquake insurance regime is based on the Earthquake Insurance Act enacted in 1966, and offers coverage not only for earthquakes, but also for tsunamis and volcanic eruptions.[157] Insurers who enroll in this scheme can offer direct coverage to earthquake damage as an extension of the optional P&C insurance policy. Individuals may choose to purchase earthquake insurance, but it is mandatory for insurers to supply it. The primary insurers cede 100 percent of the under-

---

[153]   Michel Cannarsa, Fabien Lafay and Olivier Moréteau, *France* in Financial Compensation for Victims of Catastrophe: A Comparative Legal Approach 96 (M. Faure and T. Hartlief eds., 2006).

[154]   Michel Cannarsa, Fabien Lafay and Olivier Moréteau, *France* in Financial Compensation for Victims of Catastrophe: A Comparative Legal Approach 101 (M. Faure and T. Hartlief eds., 2006).

[155]   Michael Faure and Liu Jing, *The Tsunami of March 2011 and the Subsequent Nuclear Incident at Fukushima: Who Compensates the Victims?*, 37 William & Mary Environmental Law and Policy Review 129 (2012).

[156]   World Forum of Catastrophe Programmes, *Natural Catastrophes Insurance Cover: A Diversity of Systems*, 86–90 (2008), available at http://www.wfcatprogrammes.com/c/document_library/get_file?folderId=13442&name=DLFE-553.pdf.

[157]   The Geneva Association, Insurers' Contributions to Disaster Reduction—A Series of Case Studies 47 (2013).

written earthquake insurance exposure to the Japanese Earthquake Reinsurance Scheme (JER).[158] Established by the Japanese government, the JER is responsible for reinsurance of household earthquake insurance with state guarantee.[159] In other words, the Japanese government works as a de facto reinsurer, because after primary insurers pay claims of earthquake losses, they will be compensated by the government through the JER.[160]

Because it is not mandatory for homeowners to purchase earthquake insurance, its penetration ratio is not very high. For example, the 1995 earthquake revealed a 9 percent penetration ratio. However, this figure increased to 23.7 percent following the 2011 Tohoku earthquake.[161]

Risk-based pricing According to the Law Concerning Earthquake Insurance, earthquake insurance applies risk-based premiums. Japan is divided into seven risk zones, and insurers set premiums based on the degree of exposure and building types. The premium of earthquake policies covering industrial risks and other non-household risks, for example, has normally been applied on an individual basis, depending on the basic estimate for the building structure (five types) and the location according to the degree of exposure (seven levels), ranging from 1.1 per thousand (minimum risk: class A building, level 1 location) to 18.6 per thousand (maximum risk: class E building, level 7 location).[162] The premium of household earthquake insurance is also determined in relation to two factors: the location of the property, and the type of construction.[163]

This system of premium differentiation is sometimes criticized as insufficient. For example, the division of zones has been criticized as extremely

---

[158]   The Geneva Association, Insurers' Contributions to Disaster Reduction—A Series of Case Studies 48 (2013).

[159]   Youbaraj Paudel, *A Comparative Study of Public–Private Catastrophe Insurance Systems: Lessons from Current Practices*, 37 The Geneva Papers on Risk and Insurance—Issues and Practice 257 (2012).

[160]   The Geneva Association, Insurers' Contributions to Disaster Reduction—A Series of Case Studies 48 (2013).

[161]   The Geneva Association, Insurers' Contributions to Disaster Reduction—A Series of Case Studies 49 (2013).

[162]   World Forum of Catastrophe Programmes, *Natural Catastrophes Insurance Cover: A Diversity of Systems* 88 (2008), available at http://www.wfcatprogrammes. com/c/document_library/get_file?folderId=13442&name=DLFE-553.pdf.

[163]   But the location factor consists of only four levels, and the type of building structure is only divided into wood or reinforced. See World Forum of Catastrophe Programmes, *Natural Catastrophes Insurance Cover: A Diversity of Systems* 93 (2008), available at http://www.wfcatprogrammes.com/c/document_library/get_fil e?folderId=13442&name=DLFE-553.pdf.

rough and crude, and the significant variation in earthquake risk between classes is not sufficiently reflected in the premium rating.[164]

Besides the earthquake insurance established by the Law Concerning Earthquake Insurance, cooperative insurers known as Kyosai provide the bulk of household coverage, including earthquake coverage. However, premiums provided by Kyosai do not vary by location and are less likely to induce mitigation incentives of policyholders.[165]

Contract design The JER makes use of deductibles. If the premium exceeds $550 per policy, this amount is the deductible; otherwise the deductible is equal to the premium of the policy. A maximum limit is also imposed: the total maximum limit for compensation by all insurers and government is $55.7 billion per earthquake.[166]

Loss prevention Under the JER regime, more loss prevention is conducted by the government than by insurers. This is because, first, the government controls large-scale construction and development projects in different seismic risk-zones; second, the coverage and market penetration of earthquake insurance is not very high—about 20 percent before the devastating 2011 earthquake—and it follows that insurers have fewer incentives to supply loss prevention services.[167]

Claim management Under the JER regime, claims are made via the insurance company, and are established in the individual insurance contract.

Refusal to insure Household earthquake insurers may not refuse to insure; the insured may choose whether or not to accept the insurance, but the insurers must provide it. Furthermore, primary insurers can cede all risks against earthquake for reinsurance to the JER (Earthquake Reinsurance Treaty "A"). The government will assume the ultimate risk.

For earthquake insurance covering business and industrial risks, insurers make exact assessments of the risks, and are very restrictive in terms, conditions, and ceilings. However, the supply of earthquake insurance is varied, and policyholders can choose from a large variety of options,

---

[164]   Michio Naoi et al., *Community Rating, Cross Subsidies and Underinsurance: Why So Many Households in Japan Do Not Purchase Earthquake Insurance*, 40 Journal of Real Estate Finance and Economics 544, 560 (2010).

[165]   Michael Faure and Liu, Jing, *The Tsunami of March 2011 and the Subsequent Nuclear Incident at Fukushima: Who Compensates the Victims?*, 37 William & Mary Environmental Law and Policy Review 129 (2012).

[166]   Youbaraj Paudel, *A Comparative Study of Public–Private Catastrophe Insurance Systems: Lessons from Current Practices*, 37 The Geneva Papers on Risk and Insurance—Issues and Practice 257 (2012).

[167]   K.Y. Lai et al., *The 2005 Ilan earthquake doublet and seismic crisis in northeastern Taiwan: Evidence for dyke intrusion associated with on-land propagation of the Okinawa trough*, 179 Geophysical Journal International 678 (2009).

including private insurers, the Kyosai, and in some cases local mutual funds.[168] This regulatory tool, therefore, has little function in Japan.

E. Turkey

Turkey is a land plagued with earthquakes, which cause two thirds of all natural catastrophic damages.[169] An important attempt to address this problem was the establishment of the Turkish Compulsory Insurance Pool (TCIP).[170] In 1999, Governmental Decree Law No. 587 on Compulsory Earthquake Insurance ("Decree Law") came into force and established the TCIP. One of the main objectives of the TCIP is to encourage risk reduction and mitigation practices of households.[171] As a market insurance mechanism, the TCIP supplies earthquake insurance to homeowners, and covers losses caused by earthquakes and earthquake-related catastrophes, such as fires, explosions, landslides, and tsunamis.[172] The Disaster Insurance Law (Law No. 6305), which sets out the regulations of the compulsory earthquake insurance system in detail, aims to prevent fraudulent claims and to increase the participation rate.[173] As of January 2015, the total number of policies issued was 6.8 million, the total premium collected was $380 million, the total paid claims was $80 million, the total payment capacity was $6 billion, and household participation rate stood at 38.9 percent.[174]

The TCIP is a public entity, but has no public sector employees. It is

---

[168] Michael Faure and Liu, Jing, *The Tsunami of March 2011 and the Subsequent Nuclear Incident at Fukushima: Who Compensates the Victims?*, 37 William & Mary Environmental Law and Policy Review 129 (2012).

[169] World Forum of Catastrophe Programmes, *Natural Catastrophes Insurance Cover: A Diversity of Systems* 163 (2008), available at http://www.wfcatprogrammes. com/c/document_library/get_file?folderId=13442&name=DLFE-553.pdf.

[170] The TCIP was formed with the cooperation of the World Bank, the Turkish Government, and the insurance sector. See World Forum of Catastrophe Programmes, *Natural Catastrophes Insurance Cover: A Diversity of Systems* 163–164 (2008), available at http://www.wfcatprogrammes.com/c/document_ library/get_file?folderId=13442&name=DLFE-553.pdf.

[171] Eugene Gurenko, Earthquake Insurance in Turkey: History of the Turkish Catastrophe Insurance Pool xii (2006).

[172] Burcak Başbuğ-Erkan and Ozlem Yilmaz, *Successes and Failures of Compulsory Risk Mitigation: Re-evaluating the Turkish Catastrophe Insurance Pool*, 39 Disasters 782 (2015).

[173] Burcak Başbuğ-Erkan and Ozlem Yilmaz, *Successes and Failures of Compulsory Risk Mitigation: Re-evaluating the Turkish Catastrophe Insurance Pool*, 39 Disasters 782 (2015).

[174] Burcak Başbuğ-Erkan and Ozlem Yilmaz, *Successes and Failures of*

administered by the TCIP Board of Directors, which consists of seven members drawn from government agencies, insurance companies, and the university. The government appoints an insurance or reinsurance company as the pool management company for the TCIP's daily operations.[175] Insurance companies conduct all the business tasks of the TCIP, including underwriting, claim management, and reinsuring, but they do not assume any risk. Moreover, when the payments of claims exceed the TCIP's capacity, the State provides contingent liquidity support.[176]

Risk-based pricing The TCIP adopts a differential risk-based pricing approach. According to Article 10 of the Decree Law, three factors are taken into account when determining the insurance premiums: location, construction type, and gross square area.[177] The premiums are divided into 15 tariff rates, according to the Turkey Seismic Zones Map, and into three different construction types. Consequently, risk-based pricing allows the TCIP to considerably reduce moral hazard and adverse selection.[178]

Contract design The TCIP provides a minimum 2 percent deductible to the sum insured in order to avoid "penny claims".[179] The TCIP, moreover, applies a maximum limit, and the sum for all construction types is NTL 110,000.[180] In addition, there are exclusions in the TCIP policies. For example, earthquake damage is excluded if the building was constructed after December 27, 1999, but without any valid construction license.[181]

Compulsory Risk Mitigation: Re-evaluating the Turkish Catastrophe Insurance Pool, 39 Disasters 782 (2015).

[175]   World Forum of Catastrophe Programmes, *Natural Catastrophes Insurance Cover: A Diversity of Systems* 165 (2008), available at http://www.wfcatprogrammes.com/c/document_library/get_file?folderId=13442&name=DLFE-553.pdf.

[176]   Eugene Gurenko, Earthquake Insurance in Turkey: History of the Turkish Catastrophe Insurance Pool xi (2006).

[177]   English translation of Governmental Decree Law No. 587 on Compulsory Earthquake Insurance as published in Official Gazette No. 23919 (December 27, 1999): "In determining the insurance premiums, the following factors are taken into account: square meter of the building, construction category and quality, geological characteristics of the plot of land on which the building is erected, earthquake risk, and similar factors".

[178]   Eugene Gurenko, Earthquake Insurance in Turkey: History of the Turkish Catastrophe Insurance Pool 35 (2006).

[179]   Eugene Gurenko, Earthquake Insurance in Turkey: History of the Turkish Catastrophe Insurance Pool 32–33 (2006).

[180]   World Forum of Catastrophe Programmes, *Natural Catastrophes Insurance Cover: A Diversity of Systems* 168 (2008), available at http://www.wfcatprogrammes.com/c/document_library/get_file?folderId=13442&name=DLFE-553.pdf.

[181]   Eugene Gurenko, Earthquake Insurance in Turkey: History of the Turkish Catastrophe Insurance Pool 51 (2006).

The TCIP also imposes construction maintenance obligations on the insured in the policies, as Article 14 stipulates:

> The owner who causes or allows the building and each independent section thereof to be altered contrary to the related design and in a way that will affect the load-bearing system, loses his entitlement to compensation in as much as the actual loss arises or increases because of such reason.

Loss prevention The TCIP was initiated as a loss prevention mechanism. It has played an important role in enhancing and monitoring the current National Building Code in Turkey,[182] and has also implemented revisions in land-use planning and other mitigation plans.[183] In addition, the TCIP pays much attention to education intended to raise public awareness to catastrophe risk. For example, the TCIP endeavors to introduce the concept of earthquake risk management and insurance in school textbooks.[184]

Claim management Homeowners whose houses are damaged as a result of an earthquake, and those who have a Compulsory Earthquake Insurance Policy, should consult TCIP or the insurance companies, or both, within 15 working days of becoming aware of the damage.[185] Meanwhile, loss adjustment is one of the most critical issues in the whole operation of the TCIP system due to its role in managing the moral hazard of policyholders. The TCIP employs loss adjusters who are already employed in property insurance companies.

Refusal to insure The TCIP can refuse to insure buildings that do not have a valid construction license or occupancy permit. It may also cancel the policy if the insureds make alterations to the building contrary to legislation within the insurance period.[186] The refusal or cancellation of coverage provides incentives for homeowners or builders to comply with construction codes, because homeowners who want to register

---

[182]   Burcak Başbuğ-Erkan and Ozlem Yilmaz, *Successes and Failures of Compulsory Risk Mitigation: Re-evaluating the Turkish Catastrophe Insurance Pool*, 39 Disasters 782 (2015).

[183]   Youbaraj Paudel, *A Comparative Study of Public–Private Catastrophe Insurance Systems: Lessons from Current Practices*, 37 The Geneva Papers on Risk and Insurance—Issues and Practice 257 (2012).

[184]   Eugene Gurenko, Earthquake Insurance in Turkey: History of the Turkish Catastrophe Insurance Pool xiii (2006).

[185]   Eugene Gurenko, Earthquake Insurance in Turkey: History of the Turkish Catastrophe Insurance Pool 59 (2006).

[186]   Eugene Gurenko, Earthquake Insurance in Turkey: History of the Turkish Catastrophe Insurance Pool 59 (2006).

any real-estate transaction, or open accounts for water and natural gas services, must present a valid earthquake insurance policy.[187]

V.   COMPARATIVE DISCUSSION

Controlling moral hazard and providing incentives to mitigate losses benefits both policyholders and insurers. Such efforts decrease risk and, hence, cost for policyholders, and enhance profits and financial solvency for insurers. In the context of climate change, it is especially important to integrate incentives to risk mitigation in catastrophe insurance and thus promote climate change adaptation.[188] Table 4.1 summarizes the overview of regulation by catastrophe insurance across the five countries that were explored in the previous section.

First, the question will be addressed to what extent the five technical tools aiming at disaster risk reduction are to a greater or lesser extent employed in the countries examined (A). Thereby the crucial question will also be asked to what extent this is encouraged or restricted as a result of public regulation. Then, a brief assessment of the effectiveness of disaster risk insurance in the five specific countries will be provided (B).

A.   The Use of Technical Tools

As Table 4.1 shows, all technical tools of private regulation are used to a greater or lesser extent in the countries examined. However, the effectiveness of these technical tools often depends upon the institutional setting, in other words on the public regulation. Consider for example the first and probably most important tool (notably to stimulate disaster risk reduction), being risk-based pricing. In the United Kingdom this was allowed and applied since the State refrained from intervention in premium setting as a result of the gentlemen's agreement. As indicated, this is no longer true under the new Flood Re model. Exactly the opposite, however, is the case in the United States where risk-based pricing is prohibited by the homeowner flood insurance affordability Act, as a result of which the premiums charged are substantially less than the actual risk. In France the government sets the premium for the Cat.Nat cover mandatorily by

---

[187]   Eugene Gurenko, Earthquake Insurance in Turkey: History of the Turkish Catastrophe Insurance Pool 24 (2006).

[188]   W.J. Botzen and J.C.J.M. van den Bergh, *Managing Natural Disaster Risks in a Changing Climate*, 8 Environmental Hazards—Human and Policy Dimensions 209 (2009).

Table 4.1   *Regulation by catastrophe insurance comparative table*

	UK	US	France	Japan	Turkey
Risk-based Pricing	Yes, and individualized. No longer under Flood Re	Partially, ¼ policies subsidized	No, flat rate	Yes, but for Kyosai + criticized	Yes. The TCI pool applies and the law provides the context.
Contract Design	Yes. Deductibles; a given limit for the whole content insurance.	Yes. Deductibles; maximum limit.	Yes. Deductibles; exclusions; a given limit for the whole property insurance policies.	Yes. Deductibles; maximum limit.	Yes. Deductibles; maximum limit; exclusions; insureds' obligation.
Loss Prevention	Yes. Engaging with government regulation; conducting catastrophe risk research	Yes. Mitigation assistance programs; risk-zoning and risk maps; building code regulations. NFIP promotes rebuilding in high-risk areas.	Yes. Risk prevention plan; mitigation fund.	Minimal. Low penetration.	Yes. Education, implementing mitigation measures. Monitoring via the Building Code.
Claim Management	Yes.	Yes, but costs higher than private insurance scheme.	Yes. Time limit.	Yes.	Yes. Time limit.
Refusal to Insure	Yes, and it works well due to de facto obligation of homeowners.	No.	No.	No for household earthquake insurance. Others yes.	Yes. It works well combined with compulsory insurance.

165

regulation, which excludes risk-based pricing. In Japan the law determines the system of risk differentiation applied in earthquake insurance, which is, according to some, insufficient as a tool to provide proper incentives for disaster risk reduction. In Turkey the law on compulsory earthquake insurance created the TCIP, which provides the context for risk-based pricing.

The same conclusion could be reached for the other technical tools that were examined. To take the last of these, the refusal to insure, as another example, again it appears that the ability of insurers to apply this tool very strongly depends upon the institutional context, in other words upon public regulation. In the United Kingdom the refusal to insure was possible, again under the then existing gentlemen's agreement with the government. But in the United States the refusal to insure is basically non-existent for the simple reason that it is not the insurers but the government that runs the risk under the NFIP. This seems to be the model the United Kingdom is now heading towards with Flood Re. The same conclusion can be reached for France, where the Cat.Nat cover is mandatorily included for every individual who purchases (voluntary) housing insurance. Exclusion of bad risks is hence impossible as a result of the regulation. And the same conclusion can be reached for Japan. Note that in three countries (the United States, France, and Japan) there is no possibility to refuse insurance and insurers are de facto able to transfer the consequences of bad risks to the government because in all three systems it is the government that either carries the risk (the United States) or provides reinsurance (France and Japan). In those systems compensation, even for hard-to-insure catastrophes, is provided as a result of government intervention, but at the same time one of the technical tools to stimulate disaster risk reduction by individuals (the refusal to insure) cannot be employed. An exception is the TCIP in Turkey, where a refusal to insure is possible.

A conclusion from this brief overview is that the possibilities for insurers to actively provide incentives for disaster risk reduction and hence play a role as private risk regulators strongly depend upon the institutional context and hence upon the nature of public regulation. It is often public regulation itself that prohibits the use of particular technical tools (such as premium differentiation). Of course, one has to be careful when drawing the policy conclusion that those interventions of public regulation jeopardize the development of technical tools aiming at disaster risk reduction by insurers and are therefore undesirable. It could theoretically be the case that in those countries where the public regulation limits the possibilities for insurers to use those technical tools either the government would itself be very active in developing tools of risk reduction (e.g., improving the dikes and levee system) and/or that the government would, via regulation, force homeowners to take risk-reduction measures. However, there is little evidence of this. It is known that politicians generally underinvest in disaster precaution measures because of limited political pay-offs.[189] There is also overwhelming evidence that governments systematically underinvest in disaster precaution as a result of this collective action problem[190] and regulation directed at homeowners forcing them to take specific precautionary measures is equally rare; that is why, as was stated in the introduction, regulation by insurers is often presented as a remedy to failing public regulation.[191] However, the above overview of the technical tools that would enable insurers to play this role precisely shows that it is often public regulation that restricts the possibilities of private insurers to impose measures aiming at disaster risk reduction.

B.   Country Comparison

Looking more broadly at the way in which the insurance systems described in the different countries provide incentives for disaster risk reduction one can conclude the following.

Until the beginning of this century the United Kingdom private flood insurance regime was considered a success story. Heavy floods after failing investments in flood protection by the government changed the picture.[192] Relying on risk-based premiums and other regulatory techniques, flood insurers attempted to mitigate and control the moral hazard of households. Moreover, "bad risks" were identified and regulated more rigorously, and houses become less marketable due to lack of insurance coverage. In 2013, due in part to political pressure, the UK government and insurers set up Flood Re to guarantee that high flood risk households could obtain affordable insurance. Insurers will charge policyholders at a premium that will be capped depending on the property's Council Tax band, and they will pass into Flood Re those high flood-risk homes.[193] With this new

---

[189]   Ben Depoorter, *Horizontal Political Externalities: The Supply and Demand of Disaster Management*, 56 Duke Law Journal 101 (2006).

[190]   Giuseppe Dari-Mattiacci and Michael Faure, *The Economics of Disaster Relief*, 37 Law & Policy180, 185 (2015).

[191]   Omri Ben-Shahar and Kyle D. Logue, *Outsourcing Regulation: How Insurance Reduces Moral Hazard*, 111 Michigan Law Review 197 (2012).

[192]   Michael Huber and Amodu Tola, *United Kingdom* in Financial Compensation for Victims of Catastrophe: A Comparative Legal Approach 294 (M. Faure and T. Hartlief eds., 2006).

[193]   Association of British Insurers (ABI), *Flood Re Explained* (2015), available at https://www.abi.org.uk/Insurance-and-savings/Topics-and-issues/Flooding/Government-and-insurance-industry-flood-agreement/Flood-Re-explained.

development high risk property owners will receive subsidized insurance cover, paid by all domestic property owners who have insurance, thus effectively redistributing from low to high risks.[194]

The UK system is now effectively more along the line of the NFIP in the United States. That system is subject to much stronger moral hazard, due to its partially risk-based premiums and less efficient claim management. It implicitly encourages people to live in flood hazard areas, and undermines the private insurance market.[195] It is doubtful whether the NFIP could assume future risk and potential losses because of the large number of people living in flood-prone areas, and the increase in climate-related extreme events. It is for that reason that the NFIP has been subject to a lot of criticism[196] and to proposals for reform. On the one hand it is proposed to reform the NFIP towards a model where premiums charged would better reflect risk;[197] on the other hand it is argued that the United States should move to a comprehensive natural disaster insurance regime in line with the French Cat.Nat model.[198]

Although the Cat.Nat System in France adopts a flat rate in catastrophe policies in consideration of solidarity, which means that a cross-subsidization of high risks by low risks may be justified on grounds of national solidarity, it does provide some incentives to mitigation through deductibles, through the municipal loss prevention plans (although their effectiveness has recently been challenged), and through claims management. More importantly, such a mandatory comprehensive catastrophe insurance regime allows insurers to play a more active role in regulation of individuals' behaviors than in voluntary regimes. The French model is followed by other countries, such as Belgium, where, since 2005, flooding,

---

[194]   J. Davy, *Flood Re Risk Classification and Distortion of the Market* in Future Directions of Consumer Flood Insurance in the UK. Reflections upon the Creation of Flood Re 28 (J. Hjalmarsson ed., 2015).

[195]   David Crichton, *UK and Global Insurance Responses to Flood Hazard*, 27 Water International 119 (2002).

[196]   For example, "Most homeowners remain uninsured for flood losses and the insurance that is available to cover flood losses is inadequate". See Christopher C. French, *Insuring Floods: The Most Common and Devastating Natural Catastrophes in America*, 60 Villanova Law Review 53 (2015).

[197]   Erwann Michel-Kerjan, Jeffrey Czajkowski and Howard Kunreuther, *Could Flood Insurance be Privatised in the United States? A Primer*, 40 The Geneva Papers on Risk and Insurance—Issues and Practice 179 (2015).

[198]   Howard Kunreuther, *Has the Time Come for Comprehensive Natural Disaster Insurance?* in On Risk and Disaster: Lessons from Hurricane Katrina 175 (Ronald J. Daniels et al. eds., 2006).

earthquakes, and other natural disasters are mandatorily included in all fire insurance policies.[199]

Risk-based pricing (except for Kyosai) is undoubtedly a positive aspect of the JER, and induces policyholders to take mitigation measures. However, the insurers' role is limited because of the low penetration rate (20–25 percent) of earthquake insurance for households. Given Japan's vulnerability to serious earthquakes, there seems to be a strong argument in favor of mandatory earthquake coverage, similar to the French model.

Besides its role in developed countries, catastrophe insurance becomes an increasingly important form of regulation beyond the State in many developing countries. The application of the above regulatory tools in the TCIP affirms Turkey's image as a good example and a model solution for developing and middle-income countries.[200]

VI. EXPANDING THE ROLE OF REGULATION BY CATASTROPHE INSURANCE IN CHINA

A. Regulation by Catastrophe Insurance in China

The current mechanism for managing catastrophe risks in China is known as the "Whole-Nation System" (*Juguo tizhi*), which generally refers to the government's efforts to deploy and allocate the whole nation's resources to fulfill a specific and difficult task within a limited timeframe, and thus promote the nation's interest.[201] Under the "Whole-Nation System", the government is committed to restoring social and economic order following a disaster. However, such government aid causes moral hazard, and creates negative incentives to individuals who historically rely on governmental bailouts in the wake of a catastrophe. For example, some pure forms of government bailout, including ad hoc direct payment and compensation funds, provide insufficient incentives to prevent risk and mitigate losses.[202]

---

[199]   Véronique Bruggeman, Compensating Catastrophe Victims: A Comparative Law and Economics Approach 496 (2010); Véronique Bruggeman, Michael G. Faure and Karine Fiore, *The Government as Reinsurer of Catastrophe Risks?*, 35 The Geneva Papers on Risk and Insurance—Issues and Practice 369 (2010).

[200]   J. Bommer et al., *Development of an Earthquake Loss Model for Turkish Catastrophe Insurance*, 6(3) Journal of Seismology 431 (2002).

[201]   Peijun Shi and Xin Zhang, *Chinese Mechanism against Catastrophe Risk— The Experience of Great Sichuan Earthquake*, 28 Journal of Tsinghua University (Philosophy and Social Sciences) 96 (2013).

[202]   Peijun Shi and Xin Zhang, *Chinese Mechanism against Catastrophe Risk—*

To some extent, more government bailouts may contribute to more disaster losses, because people are more likely to rely on the government to bail them out than to take precautionary measures themselves.[203] According to an empirical study on property and causality insurance in five Chinese provinces, there is a negative correlation between the amount of government relief and residents' investment in prevention measures, such as purchasing insurance.[204] Many residents admit that they are exposed to catastrophe risks, but they seldom transfer risks through insurance because they believe that the government will bail them out when catastrophes happen.[205]

Homeowner insurance is one of the least developed lines in China, and its penetration rate is low. According to a survey using face-to-face interviews, only 4 percent of interviewees had bought homeowner insurance.[206] However, people's perception of catastrophe risk and acceptance of catastrophe insurance present a more optimistic view: most people would accept catastrophe insurance, while only 4 percent of respondents considered catastrophe insurance to be unnecessary.[207] The remainder of this section examines how catastrophe insurance might be used to supplement or even supplant State governance through the "Whole-Nation System".[208] The possibility and feasibility of regulation by catastrophe insurance in China will be explored through the examination of its regulatory techniques.

The Experience of Great Sichuan Earthquake, 28 Journal of Tsinghua University (Philosophy and Social Sciences) 96 (2013).

[203] Tom Baker, *On the Genealogy of Moral Hazard*, 75 Texas Law Review 237 (1996).

[204] L. Tian and Y. Zhang, *Influence Factors of Catastrophe Insurance Demand in China—Panel Analysis in a Case of Insurance Premium Income of Five Provinces*, 26 Journal of Wuhan University of Technology (Social Science Edition) 175 (2013).

[205] He Wang, Research on Catastrophe Risk Insurance Mechanisms 5 (2013).

[206] The survey was conducted using face-to-face interviews with randomly selected respondents on trains and at railway stations in the summer of 2009. In total, 7,459 valid questionnaires were collected. The samples covered 856 different cities and counties and represented 36 percent of all cities and counties in China. See M. Wang et al., *Are People Willing to Buy Natural Disaster Insurance in China? Risk Awareness, Insurance Acceptance, and Willingness to Pay*, 32 Risk Analysis 1717 (2012).

[207] "34.7% and 39.8% of respondents believed that disaster insurance is very important and relatively important in all measures of disaster reduction and mitigation. 21.5% has no clear comments". See M. Wang et al., *Are People Willing to Buy Natural Disaster Insurance in China? Risk Awareness, Insurance Acceptance, and Willingness to Pay*, 32 Risk Analysis 1717 (2012).

[208] It is still unknown, even for pilot programs, how catastrophe insurance plays a role. No transaction information—such as risk-setting, insurance contract design, or claim management—is disclosed in the market.

Risk-based pricing According to a field research on willingness to pay (WTP), many people are willing to pay more premiums in order to acquire full coverage of property loss in catastrophe disasters.[209] In setting these premiums, regional differences and construction types should be taken into account. As was discussed above, the UK's flood insurance program is a good example of this scheme.

Urban and rural areas should receive different treatments in the proposed catastrophe insurance system because income inequality has continued to rise since China's market-oriented reform.[210] Homeowners in rural areas are low-income, and many of them could not afford insurance. China may learn from the TCIP, in which compulsory insurance for dwellings built in rural areas is not anticipated, and risk-based pricing is only applied to registered dwellings in urban areas. In fact, in the earthquake insurance pilot program in Chuxiong, the State paid the cost of every rural community's insurance in order to guarantee coverage.[211]

Contract design According to the field research, respondents who live in poor housing conditions tend to be more aware of earthquakes, and have a strong desire for insurance.[212] High deductibles may induce people to live away from hazard-prone areas, and choose stronger construction structures for their homes. When setting deductibles of policies, construction structure, house conditions, and locations should be important considerations. These tools of contract design are a common choice in the five catastrophe insurance programs discussed above.

Loss prevention According to the field research, 24.1 percent of respondents are not willing to purchase disaster house insurance because they know very little about insurance, and do not trust insurers.[213] Education, therefore, should be emphasized in insurers' loss prevention service in

---

[209]   Interviewers aimed to obtain people's WTP based on their true beliefs and feelings, as any pre-assumed ranges of premium could have misled respondents' judgment. They therefore used open-ended questions to enquire about WTP. See M. Wang et al., *Are People Willing to Buy Natural Disaster Insurance in China? Risk Awareness, Insurance Acceptance, and Willingness to Pay*, 32 Risk Analysis 1717 (2012).

[210]   Martin King Whyte, *Soaring Income Gaps: China in Comparative Perspective*, 143 Daedalus 39 (2014).

[211]   Ling Tian et al., *Perception of Earthquake Risk: A Study of the Earthquake Insurance Pilot Area in China*, 74 Natural Hazards 1595 (2014).

[212]   Ling Tian et al., *Perception of Earthquake Risk: A Study of the Earthquake Insurance Pilot Area in China*, 74 Natural Hazards 1595 (2014).

[213]   M. Wang et al., *Are People Willing to Buy Natural Disaster Insurance in China? Risk Awareness, Insurance Acceptance, and Willingness to Pay*, 32 Risk Analysis 1717 (2012).

order to create public awareness of mitigation of catastrophe risks. In addition, if more people believe in the importance of insurance in addressing catastrophe risk, catastrophe insurance will reach a higher penetration rate, as there is a strong positive correlation between the two.[214]

Claim management Insurers in China do not perform loss adjustment and claim settlement adequately. According to the field research, 23 percent of interviewees indicated that they do not trust insurers' claims management. Afraid of getting no payment after disasters, they are not willing to purchase catastrophe insurance.[215] Insurers, therefore, should increase their transparency and efficiency in order to regain the public's trust.

Refusal to insure Concerted measures and policy are required in order for this regulatory technique to play a role in China. China could follow the example of the TCIP, which requires homeowners who want to register any real-estate transaction, or open accounts for water and natural gas services, to present a valid earthquake insurance policy, and the NFIP, which stipulates that only through acquiring flood insurance for their homes, can homeowners in the 1/100 flood zones get home mortgage credits granted or secured by federal bodies or credit agencies.

In 1998, the People's Bank of China (i.e. the Chinese Central Bank) issued the Residential Mortgage Regulation, which states that before a mortgage contract is concluded, the mortgagor is required to obtain household insurance or to relegate this task to the mortgagee (Article 25). However, in 2006, the China Banking Regulatory Commission issued a notice forbidding banks from stipulating with mandatory effect that residential mortgage insurance must be acquired.[216] Although acquiring household insurance is not related to mortgage, loans or other financial services, it is still beneficial to review the series regulations and explore the feasibility of such concerted measures to be used for the take-up of catastrophe insurance.

---

[214]   M. Wang et al., *Are People Willing to Buy Natural Disaster Insurance in China? Risk Awareness, Insurance Acceptance, and Willingness to Pay*, 32 Risk Analysis 1717 (2012).
[215]   M. Wang et al., *Are People Willing to Buy Natural Disaster Insurance in China? Risk Awareness, Insurance Acceptance, and Willingness to Pay*, 32 Risk Analysis 1717 (2012).
[216]   Johanna Hjalmarsson and Mateusz Bek, *Legislative and Regulatory Methodology and Approach: Developing Catastrophe Insurance in China* in Insurance Law in China 202 (Johanna Hjalmarsson and Dingjing Huang eds., 2015).

B. Effectiveness of Regulation by Catastrophe Insurers

There is little doubt that catastrophe insurers could influence consumers' behavior. What is less clear is how effective this influence is. Theoretically speaking, both insurers and consumers present obstacles that may limit the effectiveness of regulation by catastrophe insurance. Catastrophe insurers may be reluctant to supply coverage for several reasons: first, insufficient catastrophe data impedes insurers' efforts to identify, quantify, and estimate the chances of disasters, and to set premiums for catastrophe risks; second, China's primary insurance industry does not yet have the capacity to deal with catastrophe risks, as property insurance companies do not have the capital to fully cover disaster losses; third, there are still legal restrictions that contradict catastrophe insurers' role in regulations.

Consumers, on the other hand, may reject or ignore the insurers' risk management advice, or indeed have little interest in buying catastrophe insurance at all. The "Whole-Nation System" turns relying on government's compensation into a rational choice. However, due to the low-probability nature of catastrophe disasters, and the non-rational behavior of consumers, awareness of loss prevention is weak.[217] As a result, individuals' incentive to buy insurance is diminishing.

This situation is beginning to change. Recently, China began to demand of the insurance industry that it complement government actions in addressing catastrophic risk. The 2008 Great Sichuan Earthquake and many other natural disasters over the following years, such as floods and typhoons, made the central government leaders acknowledge the contribution of insurance in regulating policyholders and compensating victims. In 2013, the 3rd Plenary Session of the 18th CPC Central Committee promulgated the Decision of the Central Committee of the Communist Party of China on Some Major Issues concerning Comprehensively Deepening the Reform. Chapter III is titled "Accelerating the Improvement of the Modern Market System", and expressly states that "we will establish an insurance system for catastrophe risks". Later, in 2014, catastrophe insurance program trials were launched in Shenzhen, in the Pearl River Delta (a densely populated metropolitan area and also one of the world's most disaster-prone regions),[218]

---

[217] T. Yue et al., *The Research on the Establishment of Chinese Catastrophe Insurance and Reinsurance System*, 16 Foreign Investment in China 255 (2013).

[218] According to recent news, in July 2014, the government of Shenzhen City bought catastrophe insurance policy from PICC on behalf of the residents of the city. This catastrophe insurance framework includes three different parts: the first is the government catastrophe assistance insurance, which is bought by the Shenzhen municipal government to supply basic assistance for all residents;

and in the Chuxiong region in the southwestern province of Yunnan, known to be prone to earthquakes.[219]

With the implementation of new practices in the near future, there is a growing need to explore the effectiveness of catastrophe insurance. This exploration should be carried out by observing and interviewing catastrophe insurance personnel (such as insurers, brokers, actuaries, loss prevention specialists, and claims professionals), a cross-section of consumers through different pilot programs, regulators of catastrophe insurance, and other government officials whose work relates to the "Whole-Nation System". This will be a prodigious undertaking, but it will give researches the opportunity to apply and evaluate the regulation of catastrophe insurance in China.

VII.   CONCLUSION

The starting point for this chapter was a recent finding in the literature that insurers increasingly act as private risk regulators, substituting or complementing public regulation. The aim was to examine which technical tools insurers use to execute this task, more particularly in the important domain of insurance for natural disasters such as flooding and earthquakes. I identified five technical tools that can be employed by insurers to, on the one hand, control the moral hazard risk and, on the other hand, provide incentives for disaster risk reduction (risk-based pricing, contract design, loss prevention, claims management, and refusal to insure). In line with the literature claiming that insurers act as private regulators, I found that when these technical tools are indeed effectively applied, insurers can fulfil their task in contributing to disaster risk reduction. However, when I then examined the possibilities in specific countries (United Kingdom, France, United States, Japan, and Turkey) to apply these technical tools I noticed that the possibilities to do so in practice are often limited, precisely as a result of public regulation. Public regulation would, for example, prohibit premium differentiation (to promote affordability of insurance) or prohibit a refusal to insure (in order to guarantee equal access to catas-

---

the second is a catastrophe fund; and the third is private catastrophe insurance. See Song Gao, *Shenzhen Signed Catastrophe Insurance Agreement for the First Time*, China Insurance Daily (2014), available at http://xw.sinoins.com/2014-07/10/content_120490.htm.

[219]   *China Says Testing Catastrophe Insurance System*, Reuters (2014), available at http://www.businessinsurance.com/article/20140820/NEWS04/140829990?Allo wView=VDl3UXk1T3hDUFNCbkJiYkY1TDJaRUt0ajBRV0ErOVVHUT09#.

trophe insurance for all citizens). As a result of those restrictions following from public regulation, insurers in many legal systems often cannot adequately play their role as private risk regulators. It would of course be too early simply to conclude therefore that public regulation interventions are necessarily undesirable. However, the interesting challenge is to examine whether it is possible to combine the political desiderata (for example of providing affordable disaster insurance to all) in a model whereby insurers could still apply their technical tools aiming at disaster risk reduction.[220] That would allow insurers to continue to play an important role as private regulators, thus substituting or complementing public regulation aimed at disaster risk reduction.

The chapter mostly focused on the question of how tools to control moral hazard in catastrophe insurance are implemented in five countries. Another equally interesting question is why the countries examined show such a variance in the implementation of tools to control moral hazard. Analysing that question is beyond the scope of this chapter but could undoubtedly be an interesting point for further research.

---

[220] Howard Kunreuther, *Long-Term Contracts for Reducing Losses from Future Catastrophes*, in Learning from Catastrophes: Strategies for Reaction and Response 235 (Howard Kunreuther eds., 2010); Howard Kunreuther, *Reflections and Guiding Principles for Dealing with Societal Risks*, in The Irrational Economist. Making Decisions in a Dangerous World 263 (E. Michel-Kerjan and P. Slovic eds., 2010).

5. Regulation by government-sponsored reinsurance in catastrophe management

I. INTRODUCTION

Reinsurance offers coverage and back-up for primary insurers. Insurers have an increasing demand for more financial capacity when underwriting catastrophic risks. The Cologne Reinsurance Company was the first professional reinsurance company, founded in 1842 following a catastrophic fire in Hamburg the same year.[1] For over a century, reinsurance has been the preferred vehicle to shed primary insurers' catastrophe risk exposure.[2] For example, reinsurers paid primary insurers 60 percent of the insured losses from the September 11 terrorist attacks, 65 percent from Hurricane Katrina, and 40 percent from Hurricane Sandy more recently.[3]

With respect to catastrophic risks, the role of reinsurance takes several forms. Reinsurance can assume a significant portion of the insured losses from primary insurers, diversify catastrophe risks globally, supply underwriting assistance, and regulate insurers' behavior to promote risk mitigation.[4] These roles often go beyond risk transfer and risk financing

---

[1] Swiss Re, *An Introduction to Reinsurance* (2002), available at http://fa2f2. voila.net/intro_reinsurance.pdf.

[2] Rajna Gibson, Michel A. Habib and Alexandre Ziegler, *Financial Markets, Reinsurance, and the Bearing of Natural Catastrophe Risk*, Swiss Finance Institute, University of Zurich (2007), available at http://www.abdn.ac.uk/business/uploads/ files/Financial%20Markets,%20Reinsurance,%20and%20the%20Bearing%20of% 20Natural%20Catastrophe%20Risk.pdf.

[3] Federal Insurance Office, *The Breadth and Scope of the Global Reinsurance Market and the Critical Role Such Market Plays in Supporting Insurance in the United States* (2014), available at http://www.treasury.gov/initiatives/fio/reports-and-notices/Documents/FIO%20-%20Reinsurance%20Report.pdf.

[4] Marcos Antonio Mendoza, *Reinsurance as Governance: Governmental Risk Management Pools as a Case Study in the Governance Role Played by Reinsurance Institutions*, 21 Connecticut Insurance Law Journal 53 (2014); Aviva Abramovsky, *Reinsurance: The Silent Regulator?*, 15 Connecticut Insurance Law Journal 345 (2009); Véronique Bruggeman, Michael Faure and Tobias Heldt,

and expand to risk regulation to primary insurers. The former role has been discussed at length in the law and economics literature,[5] but regulation by reinsurance has not been widely discussed[6] and has even been viewed as problematic. Moreover, private reinsurance has come under scrutiny due to catastrophe insurance cycles that may lead to insurance unavailability and excessive prices, especially after a major event.[7]

Government-sponsored reinsurance, which marries the merits of both government and private reinsurance, has gained increasing attention in the law and economics literature, and these programs have increased substantially in practice. Many countries use government-sponsored reinsurance to address catastrophe risks, including France (Caisse Centrale de Réassurance), Australia (Australian Reinsurance Pool Corporation), Japan (Japan Earthquake Reinsurance Co, Ltd), Turkey (Turkish Catastrophe Insurance Pool), Netherlands (Nederlandse Herverzekeringsmaatschappij voor Terrorismeschaden), Thailand (National Catastrophe Insurance Fund), United States (examples include the Terrorism Risk Insurance

---

Insurance Against Catastrophe: Government Stimulation of Insurance Markets for Catastrophic Events, 23 Duke Environmental Law and Policy Forum 185 (2012); Véronique Bruggeman, Compensating Catastrophe Victims: A Comparative Law and Economics Approach 130 (2010); David M. Cutler and Richard J. Zeckhauser, *Reinsurance for Catastrophes and Cataclysms* in The Financing of Catastrophe Risk 254 (Kenneth A. Froot ed., 1999); Federal Insurance Office, *The Breadth and Scope of the Global Reinsurance Market and the Critical Role Such Market Plays in Supporting Insurance in the United States* (2014), available at http://www.treasury.gov/initiatives/fio/reports-and-notices/Documents/FIO%20 -%20Reinsurance%20Report.pdf.

   [5]   Many articles discuss reinsurance as risk transfer and compensation to victims of catastrophes: see Véronique Bruggeman, Michael G. Faure and Karine Fiore, *The Government as Reinsurer of Catastrophe Risks?*, 35 The Geneva Papers on Risk and Insurance—Issues and Practice 369 (2010); David Durbin, *Managing Natural Catastrophe Risks: The Structure and Dynamics of Reinsurance*, 26 The Geneva Papers on Risk and Insurance—Issues and Practice 297 (2001); David M. Cutler and Richard J. Zeckhauser, *Reinsurance for Catastrophes and Cataclysms* in The Financing of Catastrophe Risk 233 (Kenneth A. Froot ed., 1999); David Cummins, *Reinsurance for Natural and Man-Made Catastrophes in the United States: Current State of the Market and Regulatory Reforms*, 10 Risk Management and Insurance Review 179 (2007).

   [6]   But see Marcos Antonio Mendoza, *Reinsurance as Governance: Governmental Risk Management Pools as a Case Study in the Governance Role Played by Reinsurance Institutions*, 21 Connecticut Insurance Law Journal 53 (2014); Aviva Abramovsky, *Reinsurance: The Silent Regulator?*, 15 Connecticut Insurance Law Journal 345 (2009).

   [7]   David Durbin, *Managing Natural Catastrophe Risks: The Structure and Dynamics of Reinsurance*, 26 The Geneva Papers on Risk and Insurance—Issues and Practice 297 (2001).

Program and the Florida Hurricane Catastrophe Fund), Belgium (Caisse nationale des calamités and the Terrorism Reinsurance and Insurance Pool), and Denmark (Terrorism Insurance Pool for Non-Life Insurance). Most of the reinsurance programs cover natural disasters and terrorism.

Meanwhile, a lot of questions about those government-sponsored reinsurance programs have been raised. One question is why the government adopts reinsurance as an intervention tool for catastrophic risks. The question more particularly arises why the government might be motivated to structure its financial support in this manner rather than in another way, such as providing direct compensation to victims of catastrophes. The question also arises whether the reinsurance industry could control the behavior of primary insurers in the same way as the primary insurers control the behavior of their insured. A related issue is of course whether the government-sponsored reinsurance programs have effectively worked in practice or whether they have resulted in some unintended consequences.

To discuss all these questions is not possible within the scope of this chapter, which will mainly argue why the Chinese government should adopt government-sponsored reinsurance and how to expand regulation by reinsurance to achieve optimal catastrophe risk management. The chapter begins by introducing basic principles of reinsurance. Next, the main regulatory techniques of reinsurance that offer primary insurers incentives to underwrite appropriately and mitigate risk are explored. The reasons why the private reinsurance market cannot provide adequate coverage for catastrophe risks and the arguments for government-sponsored reinsurance are then discussed. Next, several typical government-sponsored reinsurance programs are examined and compared, including programs in France (Caisse Centrale de Réassurance (CCR)), Japan (Japanese Earthquake Reinsurance Scheme (JERS)), and Turkey (Turkish Catastrophe Insurance Pool (TCIP)), in which primary insurers are regulated by reinsurance. Finally, it is argued that China should adopt government-sponsored reinsurance to address catastrophe risks, and the possibility and feasibility of regulation by government-sponsored reinsurance in China is discussed.

II.   REINSURANCE BASICS

A.   Introduction of Reinsurance

Reinsurance can be understood as simply insurers' insurance. Under an insurance contract, a policyholder is protected from loss by transferring risk to an insurer; analogously, under a reinsurance contract, an insurer (the cedent or ceding company) is protected from exposure by transferring risk to a reinsurer.[8] From the demand perspective, there are many theoretical explanations for a primary insurer's decision to purchase reinsurance. For example, Hoerger, Sloan and Hassan consider that the motive for reinsuring is to avoid bankruptcy, even for an insurer that is not averse to risk (a risk-neutral insurer).[9] According to other explanations, insurers require reinsurance if they face catastrophic losses, insufficient underwriting capacity, higher loss volatility, or lower surplus-to-premium ratios, or in the course of retiring from a territory or class of business.[10]

From the supply perspective, reinsurance is available from many sources, both domestic and abroad. The providers generally include professional reinsurers, pools and syndicates, direct insurers, and government agencies, and these are not mutually exclusive.[11] For example, many direct insurers are legally empowered to sell reinsurance, and they still purchase extra reinsurance from foreign professional reinsurers.

There are two broad categories of reinsurance agreements: treaty reinsurance and facultative reinsurance. Treaty reinsurance covers broad groups of policies and binds the cedent to cede a specific portion of the risk of an entire class of business, such as all property coverage written by the cedents, to a reinsurer through one contract.[12] Compared to treaty reinsurance, facultative reinsurance is often used to cover specific and catastrophic risks[13] because facultative reinsurance allows reinsurers to engage in significant underwriting prior to placing the policy and enables

---

[8]   Federal Insurance Office, *The Breadth and Scope of the Global Reinsurance Market and the Critical Role Such Market Plays in Supporting Insurance in the United States* (2014), available at http://www.treasury.gov/initiatives/fio/reports-and-notices/Documents/FIO%20-%20Reinsurance%20Report.pdf.

[9]   The authors use their model to assess how the insurer's surplus, size, and volatility of losses affect the amount of reinsurance the primary insurer purchases. See T. Hoerger, F.A. Sloan and M. Hassan, *Loss Volatility, Bankruptcy, and Insurer Demand for Reinsurance*, 3 Journal of Risk and Uncertainty 221 (1990).

[10]   Kenneth S. Abraham, Insurance Law and Regulation: Cases and Materials 739 (2005); Patrick Brockett, Robert C. Witt and Paul R. Aird, *An Overview of Reinsurance and the Reinsurance Markets*, 9 Journal of Insurance Regulation 432 (1991); Bernard L. Webb, Connor M. Harrison and James J. Markham, Insurance Operations 2 (Vol. 2, 2nd edn, 1997).

[11]   Bernard L. Webb, *Reinsurance as a Social Tool*, 1 Issue in Insurance 403, 413–414 (1984).

[12]   Barry R. Ostrager and Mary Kay Vyskocil, Modern Reinsurance Law and Practice 2-5 to 2-7 (2nd edn, 2000).

[13]   Robert H. Jerry, Understanding Insurance Law §140[b], at 1054 (4th edn, 2007); Graydon S. Staring, The Law of Reinsurance § 2:3 at 4 (Supp. 2015).

primary insurers to spread the risks of catastrophic losses that would otherwise be beyond their underwriting capacity.[14] For example, assume that a catastrophe insurance company has underwritten $100 million of earthquake insurance in a county of California but wants to retain only $10 million of the exposures. The insurer would approach one or more reinsurers and negotiate one or more agreements to provide reinsurance for the remaining $90 million. If the insurance company underwrites the earthquake coverage for another county of California, it would need to obtain reinsurance on the second county.

B.   Reinsurance for Catastrophe Insurers

In the property-casualty market, the role of reinsurance is more apparent following catastrophes than other perils. Catastrophes have a low probability of occurrence but cause very significant human and financial losses. Insurers are reluctant to underwrite catastrophes and even exclude these risks from coverage. The general theoretical explanation for why primary insurers do not cover catastrophe losses is that losses from these events are too large and too highly correlated for insurers to bear them.[15] For primary insurers, losses from catastrophes do not satisfy the conditions of statistical independence and hence are not locally insurable.[16]

Reinsurance plays a major role in making catastrophes insurable and serves an important function as protection against the accumulation of losses from catastrophes.[17] For reinsurers, because of their ability to diver-

---

[14]   Barry Ostrager and Thomas Newman, Handbook on Insurance Coverage Disputes § 15.01(b), at 997 (12th edn, 2003).

[15]   "When losses are highly correlated, insurers' claims experience is expected to be lumpy—the presence of one claim implies a likelihood of many claims. Several years may result in no claims, but some years will have gigantic levels of claims, and the strain of being prepared for a disaster year means insurers must either charge high premiums, or face the risk of bankruptcy. The conventional wisdom is that insurers choose to exclude these risks from coverage, rather than expose themselves to the year-to-year uncertainty endemic to correlated risks". See Peter Molk, *Private Versus Public Insurance for Natural Hazards: Individual Behavior's Role in Loss Mitigation* in Risk Analysis of Natural Hazards 265 (Paolo Gardoni et al. eds., 2015). See also Robert H. Jerry, Understanding Insurance Law §140[b], at 1054 (4th edn, 2007); Kenneth S. Abraham, Insurance Law and Regulation: Cases and Materials 739 (2005).

[16]   David Cummins, *Reinsurance for Natural and Man-Made Catastrophes in the United States: Current State of the Market and Regulatory Reforms*, 10 Risk Management and Insurance Review 179 (2007).

[17]   Federal Insurance Office, *The Breadth and Scope of the Global Reinsurance Market and the Critical Role Such Market Plays in Supporting Insurance in the*

sify globally, catastrophe risks can be characterized as globally insurable.[18] For example, the risk of hurricanes in the United States is independent of the risk of earthquakes in China. This provides the economic motivation for reinsurers to aggregate catastrophe risks over geographic regions and different catastrophe lines.[19] By diversifying losses across the world, catastrophes may not impose unbearable losses on the reinsurer when compared to its overall book of business, making it possible for reinsurers to provide coverage and pay losses.[20]

While primary insurance tends to be a local business, reinsurance is a much more international business, especially for catastrophic risks.[21] For example, in 2005, Hurricane Katrina caused around $90 billion in insured property losses in the United States, of which non-US reinsurers paid approximately $59 billion.[22] Because US primary insurers can access the global reinsurance market, they are able to provide coverage and pay claims.[23] The United States is not an isolated example; reinsurers have assumed a large portion of insured natural catastrophe losses in the world. For example, in 2011, global insured catastrophe losses reached $110 billion, and reinsurers assumed more than half of that amount (Figure 5.1). The largest reinsurers are in Europe and the Caribbean and are not confined to domestic reinsurers.[24]

United States (2014), available at http://www.treasury.gov/initiatives/fio/reports-and-notices/Documents/FIO%20-%20Reinsurance%20Report.pdf.

[18] Dwight Jaffee, *Catastrophe Insurance* in Research Handbook on the Economics of Insurance Law 160 (Daniel Schwarcz and Peter Siegelman eds., 2015).

[19] Dwight Jaffee, *Catastrophe Insurance* in Research Handbook on the Economics of Insurance Law 160 (Daniel Schwarcz and Peter Siegelman eds., 2015).

[20] David Cummins, *Reinsurance for Natural and Man-Made Catastrophes in the United States: Current State of the Market and Regulatory Reforms*, 10 Risk Management and Insurance Review 179 (2007).

[21] David M. Cutler and Richard J. Zeckhauser, *Reinsurance for Catastrophes and Cataclysms* in The Financing of Catastrophe Risk 237 (Kenneth A. Froot ed., 1999).

[22] Global Reinsurance Forum, Global Reinsurance: Strengthening Disaster Risk Resilience 11 (2014).

[23] David Cummins, *Reinsurance for Natural and Man-Made Catastrophes in the United States: Current State of the Market and Regulatory Reforms*, 10 Risk Management and Insurance Review 179 (2007).

[24] Europe is the origin of reinsurance business; in Europe, the insurance tax laws do allow tax-deductible reserves against future losses. In the Caribbean, a number of countries have created special tax havens. See Dwight Jaffee, *Catastrophe Insurance* in Research Handbook on the Economics of Insurance Law 160 (Daniel Schwarcz and Peter Siegelman eds., 2015).

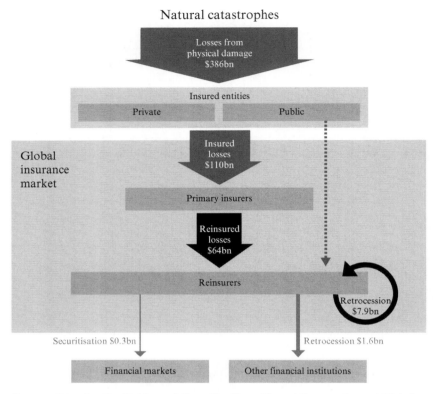

Source: Sebastian Von Dahlen and Goetz Von Peter, Natural Catastrophes and Global Reinsurance—Exploring the Linkages, BIS Quarterly Review 23 (December 2012), available at http://www.bis.org/publ/qtrpdf/r_qt1212e.pdf.

Figure 5.1   Catastrophe risk transfer in the international reinsurance market (2011)

In addition, reinsurers have developed new products such as catastrophe bonds, catastrophe derivatives, contingent capital, sidecars, and other hybrid products to facilitate new capital flows from the capital market into the reinsurance market.[25] As a result, capital in the reinsurance market

---

[25]   Catastrophe bonds are risked-linked securities that transfer catastrophe risks from insurers to investors through fully-collateralized special purpose vehicles (SPVs). Catastrophe derivatives are financial contracts used to spread catastrophe risk to capital market investors that derive value from financial

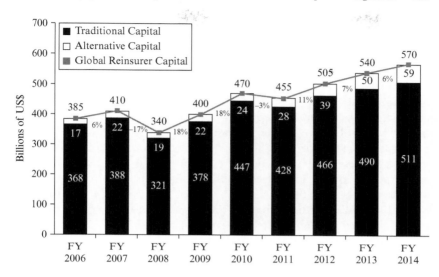

Source: Aon Benfield, The Aon Benfield Aggregate 3 (June 30, 2014), available at http:// thoughtleadership.aonbenfield.com/Documents/201409_aba_1h_2014.pdf.

Figure 5.2   Global reinsurer capital, 2006–2014

has generally been increasing year-on-year for most of the past decade (Figure 5.2).[26] For example, as of mid-2014, global reinsurance capital amounted to $570 billion ($511 billion is classified as traditional capital and $59 billion as alternative capital).[27] This accessible outside capital enables reinsurers to assume more insured catastrophe losses.

---

instruments, events, or conditions; for example, the event can be a wind storm making landfall within a certain distance of a given location. A contingent capital arrangement is a type of financing that is arranged before a loss occurs. Sidecars are SPVs formed by insurance and reinsurance companies to provide additional capacity to write reinsurance, usually for property catastrophes and marine risks. See Partner Re, *A Balanced Discussion on Insurance Linked-Securities* (2008), available at http://www.parterre.com; David Cummins, *Reinsurance for Natural and Man-Made Catastrophes in the United States: Current State of the Market and Regulatory Reforms*, 10 Risk Management and Insurance Review 179 (2007). More details are discussed in Chapter 6 of this book.

[26]   David Cummins, *Reinsurance for Natural and Man-Made Catastrophes in the United States: Current State of the Market and Regulatory Reforms*, 10 Risk Management and Insurance Review 179 (2007).

[27]   Alternative capital is the new capital that is "being channeled to specialist fund managers, who then deploy it into the insurance-linked securities (ILS) sector via products such as catastrophe bonds and industry loss warranties, or other

III.   REGULATORY TECHNIQUES OF REINSURANCE

In many respects, reinsurance often goes beyond pure risk transfer and expands to help solve catastrophic risk management issues by serving as an enforcer of compliance with government regulations and reinsurance contracts.[28] A major challenge for catastrophe reinsurance is moral hazard, a problem also encountered by primary insurance vis-à-vis policyholders. It is logical for primary insurers to change their behavior as soon as the risk is fully ceded to the reinsurer. As a private regulator, reinsurance provides incentives for the primary insurers to mitigate and prevent catastrophe losses, and thus reduce moral hazard. Reinsurance has a direct and significant impact on the business operation of primary insurance and even an indirect impact on the insureds, from contract design such as pricing, through underwriting and issuing of a policy, and ending with agreeing or refusing to pay for a claim.[29]

This section introduces four main tools that almost all reinsurers use to one degree or another to control moral hazard: loss-sensitive premiums, the duty of utmost good faith, providing risk management service, and indirect regulation of insureds. To be clear, I do not contend that these activities will exclusively solve moral hazard, nor do I contend that moral hazard management provides an adequate solution for addressing catastrophe risk. However, by supplying both the incentive and the know-how that primary insurers often lack, reinsurance can benefit both insurers and reinsurers.

A.   Loss-sensitive Premiums

Catastrophes usually cause numerous claims at the same time. Insurers tend to pass on the cost of settlements to their reinsurers, and thus the

---

'alternative' structures such as sidecars and collateralized reinsurance". See Aon Benfield, *The Aon Benfield Aggregate* 3, 20 (June 30, 2014), available at http://thoughtleadership.aonbenfield.com/Documents/201409_aba_1h_2014.pdf.

[28]   Guido Funke, *The Munich Re View on Climate-Change Litigation* in Liability for Climate Change? Experts' Views on a Potential Emerging Risk 22 (Munich Re edited, 2010); Lawrence Samplatsky, The Role of Reinsurance in Life Insurance Industry, LLM master thesis of University of Connecticut (2003).

[29]   Marcos Antonio Mendoza, *Reinsurance as Governance: Governmental Risk Management Pools as a Case Study in the Governance Role Played by Reinsurance Institutions*, 21 Connecticut Insurance Law Journal 53; Aviva Abramovsky, *Reinsurance: The Silent Regulator?*, 15 Connecticut Insurance Law Journal 345 (2009); Lawrence Samplatsky, The Role of Reinsurance in Life Insurance Industry, LLM master thesis of University of Connecticut (2003).

moral hazard problem becomes severe.[30] Traditionally, reinsurers could control moral hazard by monitoring primary insurers' business operations, including their underwriting activities and claims settlements. More importantly, reinsurers could use loss-sensitive premiums to control moral hazard. Loss-sensitive premiums generally refer to the situation where "the price of reinsurance is sensitive to concurrent reinsurance losses and to the prior period's losses total and reinsured losses".[31] Loss-sensitive premiums[32] require that reinsurance premiums should reflect an actuarially fair cost and integrate general techniques like deductibles, co-payments, and "*ex post* settling up".[33] Neil Doherty and Kent Smetters have proved that reinsurers can control for moral hazard effectively by using loss-sensitive premiums when the insurers and reinsurers are not affiliates (i.e., not part of the same financial group).[34] They present a multi-period principal-agent model of the reinsurance transaction and test it empirically. They find strong evidence for the use of loss-sensitive premiums when the insurer and reinsurer are not affiliates, and their results show that price controls can limit moral hazard.[35] Since insurers and reinsurers are generally not affiliates in underwriting catastrophe risks,[36] using loss-sensitive premiums is an effective regulatory tool for reinsurers to control moral hazard.

Is using loss-sensitive premiums feasible in practice? The answer could be yes, thanks to risk-sharing mechanisms developed by reinsurance and less rate regulation in reinsurance transactions. First, several effective

---

[30] Neil Doherty and Kent Smetters, *Moral Hazard in Reinsurance Markets*, 72 Journal of Risk and Insurance 375 (2005).

[31] Neil Doherty and Kent Smetters, *Moral Hazard in Reinsurance Markets*, 72 Journal of Risk and Insurance 375 (2005).

[32] A loss-sensitive premium is also called an actuarially fair premium, or risk-based pricing. See David M. Cutler and Richard J. Zeckhauser, *Reinsurance for Catastrophes and Cataclysms* in The Financing of Catastrophe Risk 260 (Kenneth A. Froot ed., 1999).

[33] "*Ex post* settling up" is "a retrospective adjustment of the premium based on losses incurred during the policy period that is also known as 'retrospective rating'". See Neil Doherty and Kent Smetters, *Moral Hazard in Reinsurance Markets*, 72 Journal of Risk and Insurance 375, 375–376 (2005).

[34] Neil Doherty and Kent Smetters, *Moral Hazard in Reinsurance Markets*, 72 Journal of Risk and Insurance 375 (2005).

[35] Neil Doherty and Kent Smetters, *Moral Hazard in Reinsurance Markets*, 72 Journal of Risk and Insurance 375 (2005).

[36] "Insurance of natural catastrophes is often undertaken by regional or national primary insurers and reinsured by national or international reinsurance firms." See Neil Doherty and Kent Smetters, *Moral Hazard in Reinsurance Markets*, 72 Journal of Risk and Insurance 375 (2005).

risk-sharing mechanisms are often introduced for catastrophe reinsurance premium design. The first mechanism is retrospective rating, which adjusts premiums based on losses incurred during the policy period.[37] The second is experience rating, which adjusts premiums based on losses in previous periods and which is useful when retrospective rating is not available.[38] Furthermore, because catastrophe perils are relatively rare, when series data on losses and claims are missing, the alternative method is using exposure-based modeling, which relies on scientific information and expert opinion; claims experience is only used to check and calibrate the model.[39] Second, compared to primary insurance, reinsurance markets are unregulated or very lightly regulated except in a few countries such as the United Kingdom, where reinsurers are regulated in the same way as direct insurers.[40] Lighter regulation of reinsurance enables reinsurers more flexibility to set loss-sensitive premiums.

B.   Duty of Utmost Good Faith

Primary insurers' duty of utmost good faith is the core principle of the reinsurance relationship.[41] Utmost good faith is an expressive phrase borrowed from Roman law, *uberrima fides*, which is defined as the "most abundant good faith; absolute and perfect candor or openness and honesty; the absence of any concealment or deception, however slight".[42] The reinsurance premium is less than the primary insurance premium; otherwise, primary insurers would have no incentives to underwrite such risk. Thus, reinsurers cannot duplicate the costly but necessary efforts of the primary insurer in evaluating risks and handling claims. By obligating primary insurers to act in good faith, reinsurers can control moral hazard through "invisible" monitoring without high cost.[43]

---

[37]   Neil Doherty and Kent Smetters, *Moral Hazard in Reinsurance Markets*, 72 Journal of Risk and Insurance 375 (2005).

[38]   Neil Doherty and Kent Smetters, *Moral Hazard in Reinsurance Markets*, 72 Journal of Risk and Insurance 375 (2005).

[39]   Swiss Re, *Understanding Reinsurance* 12 (2005), available at http://www.gra hambishop.com/DocumentStore/SwissRe%20Understanding%20reinsurance.pdf.

[40]   David Cummins, *Reinsurance for Natural and Man-Made Catastrophes in the United States: Current State of the Market and Regulatory Reforms*, 10 Risk Management and Insurance Review 179 (2007).

[41]   Barry R. Ostrager and Mary Kay Vyskocil, Modern Reinsurance Law 91 (2014).

[42]   Black's Law Dictionary 1520 (6th edn, 1990).

[43]   Utmost good faith as the "invisible" monitoring is borrowed from the metaphor of "the invisible hand" used by Adam Smith in economics.

The duty of utmost good faith requires the primary insurer to disclose all material facts that may affect the subject risk.[44] Those material facts may include the reinsured's underwriting process; the reinsured's amendment, renewal, or commutation in the placing of reinsurance; the payment of claims; and whether risks have been ceded fraudulently contrary to a treaty or representations.[45] As one court has stated, "[I]nsurance authorities are agreed that a ceding company, which is in possession of all the details relating to the risk, is required to exercise the utmost good faith in all its dealings with the reinsurer".[46] This places the reinsurer in the same position as the reinsured "to give him the same means and opportunity of judging ... the value of risks".[47] Utmost good faith requires the insurer to provide timely notice of claim in some courts,[48] because it permits the reinsurer "to reserve properly, to adjust premiums to reflect the loss experience under the reinsurance contract, and to decide whether to exercise the option of becoming associated with the ceding insurer in the handling and disposition of the claim".[49]

As the core principle of the reinsurance relationship, the duty of utmost good faith is enforced by many mechanisms. The first mechanism is the specific reinsurance contract provisions. This is a kind of private legislation since the parties to the reinsurance contract are professional business people. For example, reinsurers often include the "audit and inspection clauses" in the reinsurance contract, which require "the reinsured's records relative to the contract sessions to be always open to the reinsurer at reasonable times".[50] Such clauses guarantee and protect reinsurers' access to their reinsureds' underwriting and claims-handling practices. The second mechanism is court enforcement. Courts often recognize that a primary insurer's failure to act in utmost good faith offers the reinsurer a defense to its reinsurance obligation.[51] More importantly, the court requires performance of the duty by primary insurers as a condition

---

[44]   Steven Plitt et al., 1A Couch on Insurance § 9:17, at 56–57 (3rd edn, 2008).
[45]   Graydon S. Staring, The Law of Reinsurance § 8:4, at 1 (Supp. 2015).
[46]   *Northwestern Mut. Fire Ass'n v Union Mut. Fire Ins. Co of Providence*, 144 F.2d 274, 276 (9th Cir. 1944) (requiring disclosure of all material facts).
[47]   *Sun Mut. Ins. Co. v Ocean Ins. Co*, 107 U.S. 485, 510 (1883).
[48]   For example, *Fortress Re, Inc v Jefferson Insurance Co*, 465 F. Supp. 333 (E.D.N.C. 1978), aff'd 628 F.2d 860 (4th Cir. 1980); *Liberty Mutual Insurance Co v Gibbs*, 773 F.2d 15 (1st Cir. 1985).
[49]   Barry R. Ostrager and Thomas R. Newman, Handbook on Insurance Coverage Disputes § 14.02, at 466 (4th edn, 1991).
[50]   Graydon S. Staring, The Law of Reinsurance §15:8 (Supp. 2015).
[51]   See e.g. *Liquidation of Union Indemn. Ins. Co v Am. Centennial Ins. Co*, 674 N.E.2d 313, 319–320 (N.Y. 1996).

precedent to reinsurers' performance of indemnity obligation.[52] In the case of catastrophes in which reinsurance is triggered by extremely large dollar-value claims, primary insurers are more likely to take the enforcement of utmost good faith seriously. A third mechanism by which reinsurance promotes efficiency is longer-term relationship controls.[53] Reinsurance is generally not a one-off deal but conducted as a long-term relationship. Long-term relationships bond both parties, and the reinsurer can increase the effectiveness of its monitoring because it can use past experience to set future prices and terms, or even to refuse to underwrite.[54]

C.   Providing Risk Management Services

Reinsurers can act not only as capital suppliers but also as risk management service providers. For relatively simple products, reinsurers may simply act as capital suppliers. For complex products, such as underwriting catastrophic risks, reinsurers may take a more active role, more analogous to product-design consultants, through facultative reinsurance.[55] Since reinsurers deal with different catastrophe lines among geographic regions in the world, they are in a better position to share their experiences with the ceding companies. Providing risk management services for the primary insurers can take several forms:

(1)   *Entry into the market* Global reinsurers can help potential new market participants remove entry barriers, especially for those in developing countries, and allow insurers to enter this market gradually by initially reinsuring a large portion of their risks.[56]

(2)   *Product design and underwriting assistance* Reinsurers can provide expert knowledge to new market participants, and related data to

---

[52]   See e.g. *Unigard Sec. Ins. Co Inc v North River Ins. Co*, 4 F.3d 1049, 1054 (2d Cir. 1993).

[53]   This draws on the game theoretic literature in economics of repeat player dynamics, in which repeat players are less likely to engage in rent-seeking (exploitative behavior) because they want to do repeat business with the other party in the future. This idea is inspired by Professor Pat McCoy, Liberty Mutual Insurance Professor of Law, Boston College.

[54]   Aviva Abramovsky, *Reinsurance: The Silent Regulator?*, 15 Connecticut Insurance Law Journal 345 (2009).

[55]   Lawrence Samplatsky, The Role of Reinsurance in Life Insurance Industry, LLM master thesis of University of Connecticut (2003).

[56]   Patrick Brockett, Robert C. Witt and Paul R. Aird, *An Overview of Reinsurance and the Reinsurance Markets*, 9 Journal of Insurance Regulation 432 (1991).

develop a pricing model for a new product.[57] For example, from 1998 to 2002, Swiss Re, cooperating with Beijing Normal University, completed the Digital Map of China Catastrophe Events, which includes historical data on geography, weather, and so on, from the twelfth century onwards.[58] This digital map has been very helpful for the pricing of catastrophe insurance.

(3)   *Claims processing* Reinsurers can review the basis of insurers' decisions, and reinsurance contracts allow the reinsurer to opt out of an insurer's decision to deny coverage. The judgment of a reinsurer typically provides guidance to ceding insurers, which can prevent violations of unfair claims practices acts.[59]

D.   Indirect Regulation of Insureds

As well as regulating primary insurers, reinsurers may also regulate the behavior of insureds and control their moral hazard.[60] Generally speaking, reinsurers have no direct contract relationship with insureds. Because reinsurers and insureds are parties to a secondary indemnity agreement, reinsurers do not usually pay the original insureds.[61] However, under a fronting agreement arrangement,[62] the reinsurer may

---

[57]   Lawrence Samplatsky, The Role of Reinsurance in Life Insurance Industry, LLM master thesis of University of Connecticut (2003).

[58]   Xi Guo and Xinjiang Wei, *The Difficulties and Solutions for Issuing Catastrophe Bonds in China*, 6 China Insurance 22 (2005).

[59]   Lawrence Samplatsky, The Role of Reinsurance in Life Insurance Industry, LLM master thesis of University of Connecticut (2003).

[60]   The reinsurer has strong incentives to regulate insureds. Some primary insurance policies include "cut-out" provisions that allow direct action by insureds against the reinsurer. "Cut-out" provisions allow "an endorsement to an insurance policy or reinsurance contract which provides that, in the event of the insolvency of the insurance company, the amount of any loss which would have been recovered from the reinsurer by the insurance company (or its statutory receiver) will be paid instead directly to the policyholder, claimant, or other payee, as specified by the endorsement, by the reinsurer". See Reinsurance Association of America, *Fundamentals of Property and Casualty Reinsurance* 27 (2009), available at http://www.reinsurance.org/files/public/07Fundamentalsand Glossary1.pdf.

[61]   New Appleman Insurance Law Practice Guide: Separate Lines of Reinsurance § 40.01 (2007).

[62]   Despite the slightly pejorative terms used in this arrangement, there is nothing illegal in a domestic insurer acting as a front for the unauthorized insurer. In fact, so long as all other regulatory goals are met, these relationships can allow for a significant increase in insurance capacity. See New Appleman Insurance Law Practice Guide: Separate Lines of Reinsurance § 40.04 [5] (2007).

have the opportunity to regulate insureds, even indirectly. The main purpose of a fronting agreement is to allow a reinsurer that is not locally licensed to do business.[63] One New York court described a fronting agreement as an arrangement where an insurer issues a policy on a risk "with an understanding that another party will insure it".[64] Therefore, the risks underwritten by a primary insurer that has made a fronting agreement with a reinsurer will be assumed in the end by the reinsurer.[65] In other words, the reinsurer will be responsible for the entire amount that it is required to pay under the original policy. Generally, the licensed insurer will receive a fee for acting as the "front",[66] while the reinsurers can act as insurers to regulate insureds through risk-based pricing, contract design, claims management, and refusal to insure, as discussed in Chapter 4.

IV. REASONS FOR GOVERNMENT-SPONSORED REINSURANCE FOR CATASTROPHES

The previous section explored the main regulatory techniques of reinsurance that control primary insurers' moral hazard and offer them incentives to underwrite approprately and to mitigate risk. This leads to the issue of how government-provided reinsurance works and how it differs from regulation by private reinsurance. Before answering these questions, a prerequisite discussion should be why the government is involved in catastrophe reinsurance instead of leaving all catastrophe reinsurance to the private market. The main rationale for governments sponsoring catastrophe insurers and acting as reinsurers of catastrophe risks is to make up for the shortfalls of private reinsurance.

Underwriting cycles show the imperfections of private reinsurance. The phenomenon of the underwriting cycle, which refers to the tendency of insurance markets to go through alternating phases of "hard" and

---

[63]   *Union Sav. Am. Life Ins. Co v North Central Life Ins. Co*, 813 F. Supp. 481, 484 (S. D. Miss. 1993).

[64]   *Allendale Mut. Ins. Co v Excess Ins. Co*, 970 F. Supp. 265, 267 n. 2 (S.D.N.Y. 1997).

[65]   *Reliance Ins. Co v Shriver, Inc.*, 224 F.3d 641, 643 (7th Cir. 2000) (describing a fronting agreement as a "well-established and perfectly legal scheme" where policies are issued by state-licensed insurance companies and then immediately reinsured to 100 percent of face value).

[66]   *Venetsanos v Zucker, Facher & Zucker*, 638 A.2d 1333, 1336 (N.J. Super. Ct App. Div. 1994).

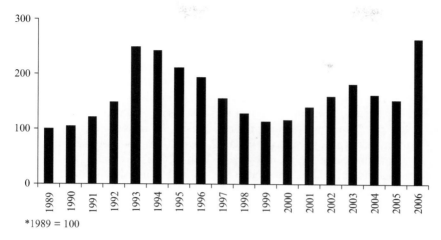

Note: The rate on line is a pricing concept, which is found by dividing the contractual reinsurance premium by the reinsurance limit and converting the result into a percentage. See Kenneth Froot, The Intermediation of Financial Risks: Evolution in the Catastrophe Reinsurance Market, 11 Risk Management and Insurance Review 281 (2008).

Figure 5.3   US catastrophe reinsurance: rate on line index

"soft" markets, is an important characteristic of insurance markets.[67] Hard markets are usually triggered by capital depletions resulting from underwriting catastrophic losses of unexpected magnitude.[68] Figure 5.3 shows the infamous cyclical nature of property-casualty insurance for the years following 1989. It clearly indicates that reinsurance prices are cyclical.[69] The hard market in the 1990s was caused by Hurricane Andrew (1992). The magnitude of losses from Andrew took insurers by surprise, and 13 insurance companies went bankrupt primarily as a result of capital depletions.[70] After the catastrophe, insurance companies improved loss

---

[67]   Hard market leads to decreased supply but increased premium, whereas in a soft market, coverage supply is plentiful and prices decline. See David Cummins and Olivier Mahul, Catastrophe Risk Financing in Developing Countries: Principles for Public Intervention 194 (2009).

[68]   David Cummins, *Reinsurance for Natural and Man-Made Catastrophes in the United States: Current State of the Market and Regulatory Reforms*, 10 Risk Management and Insurance Review 179 (2007).

[69]   Reinsurance prices increased and supply contracted following Hurricane Andrew in 1992, paralleling the market response to later 2005 hurricane seasons.

[70]   A.M. Best Company, 2006 Annual Hurricane Study: Shake, Rattle, and Roar (2006).

estimation and risk management capabilities; insurers and catastrophe modeling firms revised their expectations of future hurricane losses upward.[71] Accordingly, prices of reinsurance increased for the 1993 renewals. There was a similar price increase after Hurricane Katrina.

To some extent, reinsurers face similar financing limitations to those faced by primary insurers.[72] During periods of hard markets, there is often insufficient reinsuring capacity. Why are so few assets allocated to catastrophe reinsurance? Since market distortions appear to be more supply- (reinsurer) than demand- (primary insurer) related,[73] explanations for imperfections in the reinsurance market mainly consider supply restrictions. The explanations below are well documented in the law and economics literature.

First, informational asymmetries between capital providers and reinsurers about exposure levels and reserve adequacy can result in high costs of capital during hard markets.[74] It may be more costly for reinsurers to raise additional funds since capital providers cannot clearly separate performance into event losses and reinsurers' skill in peril selection, i.e. distinguishing good risks and bad risks and underwriting the good risks.[75] Irrational investor behavior, such as investor "trend-following", may also decrease the supply of capital to reinsurance after a major catastrophe.[76] The consensus in the economics literature is that shortages are driven by imperfections in the capital and insurance markets that prevent capital

---

[71]   David Cummins, *Reinsurance for Natural and Man-Made Catastrophes in the United States: Current State of the Market and Regulatory Reforms*, 10 Risk Management and Insurance Review 179 (2007).

[72]   Many primary insurers do not have enough capital and surplus themselves to survive catastrophes, and they have to rely upon the reinsurance market to recompense catastrophic damages. See Véronique Bruggeman, Compensating Catastrophe Victims: A Comparative Law and Economics Approach 136 (2010).

[73]   According to a set of demand–supply equilibrium points, graphed in terms of price and quantity of reinsurance provided, Froot shows a strong negative correlation between price and quantity supplied. This suggests that supply shocks are the main driver rather than demand—a decline in supply results in an increase in price and decline in quantity of risk transfer. See Kenneth Froot, *The Intermediation of Financial Risks: Evolution in the Catastrophe Reinsurance Market*, 11 Risk Management and Insurance Review 281 (2008).

[74]   David Cummins and Olivier Mahul, Catastrophe Risk Financing in Developing Countries: Principles for Public Intervention 194–195 (2009).

[75]   Kenneth Froot, *The Market for Catastrophe Risk: A Clinical Examination*, 60 Journal of Financial Economics 529 (2001).

[76]   Investor trend-following refers to the situation where investors expect recent performance to continue; as a result, they tend to buy exposures that have recently performed well and to sell those that have not. See Kenneth Froot. *The Market for Catastrophe Risk: A Clinical Examination*, 60 Journal of Financial Economics 529 (2001).

from flowing freely into and out of reinsurance corporations in response to catastrophic losses.[77]

A major catastrophe may deplete reinsurer capital and surplus, and require some time to replenish.[78] Without additional funds from capital providers, such depletion of equity capital is likely to result in raised premiums for reinsurance that are above the expected loss of coverage.[79] Using the empirical evidence during the year following Hurricane Andrew, in relation to insurers that had greater exposure to the southeastern United States and to hurricanes wherever they occur, Froot demonstrated that in terms of reinsurance "prices rise most where quantities decline most".[80]

Second, reinsurers may have market power, and supply shortages and high prices following catastrophes may occur because reinsurers have no incentive to increase their capital. By putting less money at risk and preventing new entry, incumbent reinsurers keep prices high.[81] The former Massachusetts Insurance Commissioner argued that market power among reinsurers is the main reason why catastrophe reinsurance has proved more profitable than insurance.[82] Barriers to entry are also relevant to the market power story.[83] No entry barriers would tend to suggest that there is no market power; it is entry barriers that permit sellers to keep prices above marginal costs. Froot has noted that the 1990s were not crisis years, but sellers could have been poised for entry when and if prices of reinsurance rose.[84]

---

[77]   Ralph Winter, *The Dynamics of Competitive Insurance Markets*, 3 Journal of Financial Intermediation 379 (1994); David Cummins and Patricia M. Danzon, *Price Shocks and Capital Flows in Liability Insurance*, 6 Journal of Financial Intermediation 3 (1997); David Cummins and Neil A. Doherty, *Capitalization of the Property–Liability Insurance Industry: Overview*, 21 Journal of Financial Services Research 5 (2002); David Cummins and Olivier Mahul, Catastrophe Risk Financing in Developing Countries: Principles for Public Intervention 194 (2009).

[78]   Kenneth Froot, *The Intermediation of Financial Risks: Evolution in the Catastrophe Reinsurance Market*, 11 Risk Management and Insurance Review 281 (2008).

[79]   Frank A. Sloan and Lindsey M. Chepke, *Reinsurance* in Medical Malpractice 247, 252–253 (2008).

[80]   Kenneth Froot, *The Market for Catastrophe Risk: A Clinical Examination*, 60 Journal of Financial Economics 529 (2001).

[81]   Kenneth Froot, *The Market for Catastrophe Risk: A Clinical Examination*, 60 Journal of Financial Economics 529 (2001).

[82]   Kenneth Froot, *The Market for Catastrophe Risk: A Clinical Examination*, 60 Journal of Financial Economics 529 (2001).

[83]   Kenneth Froot, *The Market for Catastrophe Risk: A Clinical Examination*, 60 Journal of Financial Economics 529 (2001).

[84]   Kenneth Froot, *The Market for Catastrophe Risk: A Clinical Examination*, 60 Journal of Financial Economics 529 (2001).

Third, the corporate form of reinsurance ownership may also contribute to short supply in the reinsurance market in the wake of catastrophes.[85] Corporations create agency costs because the interests of managers ("agents") may not perfectly align with those of shareholders ("principals"). Managers act in many ways that do not maximize corporation value but instead advance their personal financial interests.[86] Such corporate behavior may not be efficient in supplying reinsurance coverage.

In sum, due to financing limitations of private reinsurers caused by informational asymmetries, market power, and corporate forms of reinsurance ownership, the practice of reinsurance demands some kind of government intervention.

V. GOVERNMENT-SPONSORED CATASTROPHE REINSURANCE PROGRAMS: EXAMPLES

Section III above described the tools available to reinsurers to regulate insurers and underwritten catastrophe risks. We saw that through contract design (loss-sensitive premiums), the duty of utmost good faith, providing risk management services, and indirect regulation of insureds, reinsurance has the capacity to perform a social function that is regulatory in nature— less moral hazard on the part of primary insurers and better preparedness on the part of insureds. Section IV above explained why much of the reinsurance for catastrophe risks in the world is sponsored by governments. Compared with the capital shortfall of private reinsurers, governments can channel capital effectively and quickly following catastrophes since they can raise money through tax or borrow money by issuing debt or government bonds.[87]

---

[85]   Kenneth Froot, *The Intermediation of Financial Risks: Evolution in the Catastrophe Reinsurance Market*, 11 Risk Management and Insurance Review 281 (2008). Kunreuther, Pauly and Russell suggest that capital suppliers may believe that the high losses they experienced are not random, which reflects reinsurer mismanagement. See H. Kunreuther, M. Pauly and T. Russell, *Demand and Supply Side Anomalies in Catastrophe Insurance Markets: The Role of the Public and Private Sectors*, Paper prepared at the MIT/LSE/Cornell Conference on Behavioral Economics (2004).

[86]   Kenneth Froot, *The Market for Catastrophe Risk: A Clinical Examination*, 60 Journal of Financial Economics 529 (2001); Frank A. Sloan and Lindsey M. Chepke, *Reinsurance* in Medical Malpractice 247, 253 (2008).

[87]   David M. Cutler and Richard J. Zeckhauser, *Reinsurance for Catastrophes and Cataclysms* in The Financing of Catastrophe Risk 258–259 (Kenneth A. Froot ed., 1999).

This section examines how government-sponsored reinsurance programs work. Government-sponsored reinsurance is increasingly welcomed by law and economics scholarship as a way to manage catastrophic risks.[88] Meanwhile, government-sponsored reinsurance has increased substantially in practice, and many programs are established when primary-insurance markets break down. It is not possible within the scope of this chapter to critically analyse all of the programs that exist, some of which were mentioned in the introduction. Accordingly, this discussion will be limited to the French CCR, the Japanese JERS and the Turkish TCIP. As these examples demonstrate, there is wide variation in the nature and extent of regulation of catastrophe reinsurance across different countries.

Government-sponsored reinsurance is a kind of public–private partnership that combines the merits of both government and reinsurance.[89] The origins of such partnerships can be traced to the nuclear liability conventions that emerged in the 1960s.[90] Government-sponsored reinsurance programs have since expanded to many lines of insurance, including medical malpractice,[91] expropriation insurance,[92] crop insurance

---

[88] For example, Véronique Bruggeman, Michael Faure and Tobias Heldt, *Insurance Against Catastrophe: Government Stimulation of Insurance Markets for Catastrophic Events*, 23 Duke Environmental Law and Policy Forum 185 (2012); Howard Kunreuther and Erwann Michel-Kerjan, *Managing Catastrophic Risks through Redesigned Insurance: Challenges and Opportunities* in Handbook of Insurance 517 (G. Dionne ed., 2013); Véronique Bruggeman, Michael Faure and Karine Fiore, *The Government as Reinsurer of Catastrophe Risks?*, 35 The Geneva Papers on Risk and Insurance—Issues and Practice 369 (2010).

[89] Howard Kunreuther and Mark Pauly, *Rules Rather than Discretion: Lessons from Hurricane Katrina*, 33 Journal of Risk and Uncertainty 101 (2006): Saul Levmore and Kyle D. Logue, *Insuring Against Terrorism—and Crime*, 102 Michigan Law Review 268 (2003).

[90] The Price-Anderson Act, concerning nuclear facilities, is an example of this model. See Véronique Bruggeman, Michael G. Faure and Karine Fiore, *The Government as Reinsurer of Catastrophe Risks?*, 35 The Geneva Papers on Risk and Insurance—Issues and Practice 369 (2010).

[91] For example, New Jersey enacted the New Jersey Medical Malpractice Reinsurance Association in 1976, and any member of the association could be approved by the association to write malpractice coverage. The insurer would then be reinsured by the association either in full or in part. See Vincent R. Zarate, *N.J. Malpractice Unit Activated*, Journal of Commerce 9 (6 January 1977).

[92] For example, in the United States, expropriation insurance written by the Overseas Private Investment Corporation (OPIC) was a purely governmental program, and eventually OPIC turned the program over to private insurers, with OPIC functioning only as a reinsurer. See Bernard Webb, *Reinsurance as a Social Tool*, 1 Issue in Insurance 403, 448 (1984).

programs,[93] and terrorism insurance after the September 11 terrorist attack.[94] Since a government has a deep credit capacity due to its ability to raise money through tax or borrow money by issuing debt far more readily than private insurers or reinsurers,[95] it is widely recognized that a government can help address catastrophic risks in some respects, and can thus be used to support the failures of the primary insurance market.[96] In this way, government reinsurance can expand the supply of catastrophe insurance in the primary market.

A.   The French CCR

The French government-sponsored reinsurance for natural disasters takes the form of subsidized government reinsurance with mandatory private primary insurance.[97] Historically, in France, private insurers offered little coverage for natural catastrophe risks, and the government intervened through ad hoc assistance in the aftermath of disasters until 1982.[98] The 1982 disaster law required private insurers to underwrite catastrophic

---

[93]   For example, the Federal Crop Insurance Corporation (FCIC) is authorized to provide reinsurance for "all risks" crop written by private insurers. See *U.S. Code Congressional and Administrative News*, No. 9 (November 1980) 5949.

[94]   After September 11, 2001, when airline risks became more difficult to insure, the US federal government guaranteed insurance coverage. See Kenneth Abraham, *United States of America. Liability for Acts of Terrorism under US Law* in Terrorism, Tort Law and Insurance: A Comparative Survey 176 (B.A. Koch ed., 2004).

[95]   Louis Kaplow, *Incentives and Government Relief for Risk*, 4 Journal of Risk and Uncertainty 167 (1991).

[96]   For example, John V. Jacobi, *Government Reinsurance Programs and Consumer-Driven Care*, 53 Buffalo Law Review 537 (2005); Daniel A. Schenck, *Next Step for Brownfields Government Reinsurance of Environmental Cleanup Policies*, 10 Connecticut Insurance Law Journal 401 (2003); Mark A. Hall, *Government-sponsored Reinsurance*, 19 Annals of Health Law 465 (2010); Véronique Bruggeman, Michael Faure and Tobias Heldt, *Insurance Against Catastrophe: Government Stimulation of Insurance Markets for Catastrophic Events*, 23 Duke Environmental Law and Policy Forum 185 (2012); Howard Kunreuther and Erwann Michel-Kerjan, *Managing Catastrophic Risks through Redesigned Insurance: Challenges and Opportunities* in Handbook of Insurance 517 (G. Dionne ed., 2013).

[97]   Lorilee Medders, Kathleen McCullough and Verena Jäger, *Tale of Two Regions: Natural Catastrophe Insurance and Regulation in the United States and the European Union*, 30 Journal of Insurance Regulation 171, 184 (2011).

[98]   David A. Moss, *Courting Disaster? The Transformation of Federal Disaster Policy since 1803* in The Financing of Catastrophe Risk 307 (Kenneth A. Froot ed., 1999).

risks and permitted them to cede those risks to the Caisse Centrale de Réassurance (CCR), the state-guaranteed reinsurer.[99] To gain the benefit of the government guarantee, the CCR pays an annual "premium" to the government (Article R. 431-16-2 Insurance Code), similar to private retrocession.[100]

The CCR provides a coverage system that consists of two tiers of reinsurance based on two separate treaties: a 50 percent quota share treaty and a stop-loss treaty with an unlimited governmental guarantee.[101] According to the 50 percent quota share treaty, the insurer cedes 50 percent of its premiums and its risk to the CCR, which also helps to reduce anti-selection issue of primary insurers.[102] Those risks not covered by the quota share treaty are subject to the stop-loss treaty. The stop-loss treaty, with an unlimited governmental guarantee, enables primary insurers to underwrite high severity hazards.

Loss-sensitive premiums Loss-sensitive premiums require that reinsurance premiums should reflect an actuarially fair cost and reinsured losses. Thanks to an unlimited guarantee from the French Treasury, in the first 15 years the CCR could offer low prices to all ceding companies on identical terms.[103] In 1997, the CCR revised its reinsurance terms because of a deterioration in the claims figures and changes in the primary insurance market. It began to move forward to loss-sensitive premiums setting, and its rating of the "stop-loss" covers was based upon each individual insurer's loss record.[104]

---

[99] Decree No. 82-706 of 10 August 1982 on the Reinsurance Operations for the Natural Catastrophe Risks by the *Caisse Centrale de Réassurance*. Application of Article 4 of the Act No. 82-600 of 13 July 1982, JORF 11 August 1982. See Véronique Bruggeman, Michael G. Faure and Karine Fiore, *The Government as Reinsurer of Catastrophe Risks?*, 35 The Geneva Papers on Risk and Insurance—Issues and Practice 369 (2010).

[100] Suzanne Vallet, *Insuring the Uninsurable: The French Natural Catastrophe Insurance System* in Catastrophe Risk and Reinsurance: A Country Risk Management Perspective 119 (Eugene N. Gurenko ed., 2004).

[101] Lorilee Medders, Kathleen McCullough and Verena Jäger, *Tale of Two Regions: Natural Catastrophe Insurance and Regulation in the United States and the European Union*, 30 Journal of Insurance Regulation 171, 184 (2011).

[102] Lorilee Medders, Kathleen McCullough and Verena Jäger, *Tale of Two Regions: Natural Catastrophe Insurance and Regulation in the United States and the European Union*, 30 Journal of Insurance Regulation 171, 184 (2011).

[103] Suzanne Vallet, *Insuring the Uninsurable: The French Natural Catastrophe Insurance System* in Catastrophe Risk and Reinsurance: A Country Risk Management Perspective 211 (Eugene N. Gurenko ed., 2004).

[104] Such price setting does not include the quota share treaty. See Suzanne Vallet, *Insuring the Uninsurable: The French Natural Catastrophe Insurance System*

Such loss-sensitive premiums setting represents a good start, but it still has a long way to go. With the governmental guarantee, the CCR charges relatively lower premiums to primary insurers than other private reinsurance companies and thus crowds them out of the market.[105] On the other hand, it will be the taxpayers who ultimately pay the CCR's unlimited coverage that can offset damages.[106] France's relatively moderate exposure to natural disasters makes the operation of the CCR suitable for France. It is still questionable to what extent the CCR is capable of dealing with the next mega-catastrophe.

Duty of utmost good faith The duty of utmost good faith is enforced by two mechanisms in the operation of the CCR. First, the 50 percent quota share treaty contributes to primary insurers' performance of the duty of utmost good faith. Primary insurers have to retain half of the risks themselves under the 50 percent quota share treaty, which gives them an incentive to underwrite appropriately.[107] Second, the long-term relationship between the CCR and the ceding companies also contributes to the performance of the duty of utmost good faith. As the state-guaranteed reinsurer, the CCR has operated for several decades and has detailed records of the ceding companies. Such experiences help the CCR effectively to monitor primary insurers' performance of utmost good faith.

Providing risk management services It is not clear whether the CCR provides risk management services for ceding companies. Nonetheless, as one of the top 20 reinsurance carriers in the world with an AAA rating from Standard & Poor's, the CCR clearly has expertise in risk management.[108] Dealing with ceding companies of different sizes, differing legal forms, and various types of portfolios, the CCR is in a better position to share

---

in Catastrophe Risk and Reinsurance: A Country Risk Management Perspective 211–212 (Eugene N. Gurenko ed., 2004).

[105]   Erwann Michel-Kerjan, *Catastrophe Economics: The National Flood Insurance Program*, 24 Journal of Economic Perspectives 165 (2010). "The CCR is not a monopolistic disaster reinsurer. In fact, there are several reinsurers writing business with primary reinsurers in France". See Lorilee Medders, Kathleen McCullough and Verena Jäger, *Tale of Two Regions: Natural Catastrophe Insurance and Regulation in the United States and the European Union*, 30 Journal of Insurance Regulation 171, 185 (2011).

[106]   Lorilee Medders, Kathleen McCullough and Verena Jäger, *Tale of Two Regions: Natural Catastrophe Insurance and Regulation in the United States and the European Union*, 30 Journal of Insurance Regulation 171, 185 (2011).

[107]   Suzanne Vallet, *The French Experience in the Management and Compensation of Large-scale Disasters* in Catastrophic Risks and Insurance 293 (OECD, 2005).

[108]   Lorilee Medders, Kathleen McCullough and Verena Jäger, *Tale of Two Regions: Natural Catastrophe Insurance and Regulation in the United States and the European Union*, 30 Journal of Insurance Regulation 171, 185 (2011).

its experiences in managing catastrophe risk and providing coverage for multiple types of natural hazards.

Indirect regulation of insureds Since the CCR is licensed to conduct business in France, there is no need for a fronting agreement arrangement. There is no empirical evidence of the CCR's indirect regulation of insureds.

B. The Japanese JERS

The Japanese government-sponsored reinsurance for earthquakes takes the form of the government providing reinsurance capacity. The Japan Earthquake Reinsurance Co, Ltd (JERS) was established based on the Act on Earthquake Insurance in 1966 enacted after the Niigata Earthquake in 1964.[109] Primary insurers issue standard residential policies that cover losses to personal dwellings and contents caused by earthquakes and volcanic eruptions and then cede these risks to the JERS.[110] The JERS is a specialized reinsurance company but backed by the Japanese government. It can also be seen as an earthquake reinsurance pool, retaining a portion of the liability and retroceding the rest to private insurers (based on their market share) and to the Japanese government through reinsurance treaties.[111] To be clear, the JERS only covers personal residential, not commercial, earthquake insurance.

The Japanese government is not involved in professional reinsurance business operations, which are all managed by the JERS. Nevertheless, the successful operation of the JERS depends on a commitment from the Japanese government, which provides significant reinsurance capacity as a last resort.[112] The aggregate limit of indemnity for earthquake insurance liabilities (JPY 6.2 trillion) is shared by private insurers and the government at different tiers. The first tier, which covers earthquake insurance liabilities up to JPY 85 billion, is completely compensated by the JERS;

---

[109]   OECD, *Disaster Risk Financing in APEC Economies* 73 (2013), available at http://www.oecd.org/daf/fin/insurance/OECD_APEC_DisasterRiskFinancing. pdf.

[110]   Yuichi Takeda, *Government as Reinsurers of Last Resort: The Japanese Experience* in Catastrophe Risk and Reinsurance: A Country Risk Management Perspective 225 (Eugene N. Gurenko ed., 2004).

[111]   Michael Faure and Jing Liu, *The Tsunami of March 2011 and the Subsequent Nuclear Incident at Fukushima: Who Compensates the Victims?*, 37 William & Mary Environmental Law and Policy Review 129 (2012).

[112]   Yuichi Takeda, *Government as Reinsurers of Last Resort: The Japanese Experience* in Catastrophe Risk and Reinsurance: A Country Risk Management Perspective 225 (Eugene N. Gurenko ed., 2004).

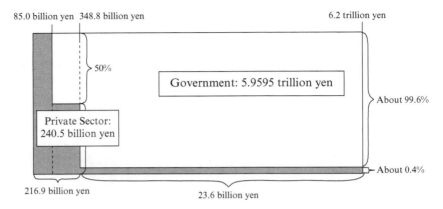

Source: a   OECD, Disaster Risk Financing in APEC Economies 73 (2013), available at http://www.oecd.org/daf/fin/insurance/OECD_APEC_DisasterRiskFinancing.pdf.

Figure 5.4   Risk allocation under the Japanese Earthquake Reinsurance Scheme

the second tier, which covers earthquake insurance liabilities over JPY 85 billion and up to JPY 348.8 billion, is compensated by equal contributions from the Japanese government (50 percent) and the JERS and private insurers (due to retroceded risk from the JERS: 50 percent); the third tier, which covers earthquake insurance liabilities from JPY 348.8 billion to JPY 6.2 trillion, is mostly compensated by the Japanese government (99.6 percent) and a very small share by private insurers (0.4 percent)[113] (see Figure 5.4). If the earthquake insurance liabilities of one catastrophe exceed JPY 6,200 billion, residential policyholders' claims are reduced proportionately following the provisions of the Act on Earthquake Insurance.[114]

Loss-sensitive premiums Making the premiums loss-sensitive is one of the most challenging tasks for a public–private partnership. The JERS is no exception. The reinsurance price of the JERS is not market-based

    [113]   Yuichi Takeda, *Government as Reinsurers of Last Resort: The Japanese Experience* in Catastrophe Risk and Reinsurance: A Country Risk Management Perspective 225 (Eugene N. Gurenko ed., 2004); OECD, *Disaster Risk Financing in APEC Economies* 73 (2013), available at http://www.oecd.org/daf/fin/insurance/ OECD_APEC_DisasterRiskFinancing.pdf.
    [114]   OECD, *Disaster Risk Financing in APEC Economies* 73 (2013), available at http://www.oecd.org/daf/fin/insurance/OECD_APEC_DisasterRiskFinancing. pdf.

but determined by the Japanese government. The premiums are not loss-sensitive but are set to follow a general fair-value principle.[115]

Duty of utmost good faith The primary insurers' duty of utmost good faith is extremely important for the JERS. The primary insurers are allowed to cede 100 percent of the underwritten earthquake insurance exposure to the JERS.[116] If primary insurers underwrite inappropriately, the JERS will assume all the bad risks. According to the requirement of utmost good faith, the primary insurers should disclose all material facts that may affect the subject risk. In order to enforce such a requirement, the Japanese government stipulates that all rating work is undertaken solely by the Non-Life-Insurance Rating Organization of Japan (NLIRO) and not by primary insurers.[117] The NLIRO has to file materials setting, modifying, and revising the base rates to the Financial Supervisory Authority for approval.[118] Under this system, the JERS is able to access the underwriting materials of its ceding companies. The duty of utmost good faith is also enforced by reinsurance treaty provisions. The Earthquake Reinsurance Treaty between the JERS and private insurance companies includes the retrocession provision, which provides that primary insurers cede their underwritten risks to the JERS, and the JERS in turn retrocedes the risks in the second layer to the primary insurers and the Japanese government in equal portion.[119] Retroceding 50 percent of the risk in the second tier to primary insurers contributes to their performance of the duty of utmost good faith.

Providing risk management service One purpose of establishing the JERS

---

[115]   Currently, the details of JER reinsurance contracts are not fully disclosed, except the names of the counterparties and the amount of reinsurance. It is difficult to explain the basic elements of the general fair-value principle. Some anecdotes from the Japanese insurance industry imply that affordability and sustainability are both important considerations of this principle. See Yuichi Takeda, *Government as Reinsurers of Last Resort: The Japanese Experience* in Catastrophe Risk and Reinsurance: A Country Risk Management Perspective 225 (Eugene N. Gurenko ed., 2004).

[116]   The Geneva Association, Insurers' Contributions to Disaster Reduction—A Series of Case Studies 48 (2013).

[117]   Yuichi Takeda, *Government as Reinsurers of Last Resort: The Japanese Experience* in Catastrophe Risk and Reinsurance: A Country Risk Management Perspective 225 (Eugene N. Gurenko ed., 2004).

[118]   Yuichi Takeda, *Government as Reinsurers of Last Resort: The Japanese Experience* in Catastrophe Risk and Reinsurance: A Country Risk Management Perspective 225 (Eugene N. Gurenko ed., 2004).

[119]   Non-Life Insurance Rating Organization of Japan and K. Kawachimaru, *Disaster Risk Management in Japan* in Catastrophic Risks and Insurance 303 (OECD, 2005).

is to facilitate loss mitigation and a recovery process through the insurance industry. However, in practice, the NLIRO, rather than the JERS, undertakes major service works for primary insurers.

Indirect regulation of insureds Since the JERS is licensed to conduct business in Japan, there is no need for a fronting agreement arrangement. The JERS has incentives to regulate insureds' behavior and awareness of earthquake risks because primary insurers cede 100 percent of the risks to the JERS. For example, the JERS uses deductibles to enhance individuals' risk mitigation efforts.[120]

C.   The Turkish TCIP

Unlike the French CCR and the Japanese JERS, the Turkish government has not established a specific reinsurance company to assume catastrophe risk. Instead, the Turkish government provides contingent liquidity support when the payments of claims exceed the capacity of the Turkish Catastrophe Insurance Pool (TCIP).[121] This could be regarded as reinsurance since it is the last resort. The first layer reinsurance arrangement under the mechanisms of the TCIP is provided by the international reinsurers, which assume the transferred risks from the TCIP. Therefore, the regulatory techniques of reinsurance include both international reinsurers and the Turkish government.

In 1999, Governmental Decree Law No. 587 on Compulsory Earthquake Insurance ("Decree Law") came into force and created the TCIP in the aftermath of the devastating Marmara Earthquake. The TCIP is a public–private partnership (Figure 5.5). Insurance companies act as agents for the TCIP and cede 100 percent of all risks acquired by the TCIP; and they receive a commission from the pool.[122] The TCIP transfers risks to international reinsurers through sharing pools under

---

[120]   If the premium exceeds $550 per policy, this amount is the deductible; otherwise the deductible is equal to the premium of the policy. See Youbaraj Paudel, *A Comparative Study of Public–Private Catastrophe Insurance Systems: Lessons from Current Practices*, 37 The Geneva Papers on Risk and Insurance—Issues and Practice 257 (2012).

[121]   Eugene Gurenko, Earthquake Insurance in Turkey: History of the Turkish Catastrophe Insurance Pool xi–xii (2006).

[122]   The TCIP provides earthquake insurance to homeowners, and covers losses caused by earthquakes and earthquake-related catastrophes, such as fires, explosions, landslides, and tsunamis. See Johann-Adrian von Lucius, *A Reinsurer's Perspective on the Turkish Catastrophe Insurance Pool (TCIP)* in Catastrophe Risk and Reinsurance: A Country Risk Management Perspective 217 (Eugene N. Gurenko ed., 2004); Burcak Başbuğ-Erkan and Ozlem Yilmaz, *Successes and*

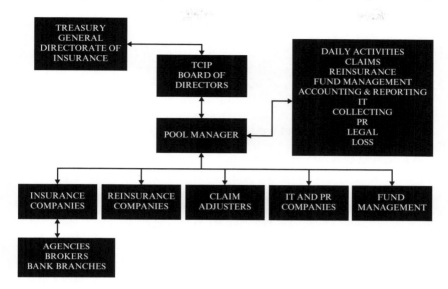

Source: Burcak Başbuğ-Erkan and Ozlem Yilmaz, Successes and Failures of Compulsory Risk Mitigation: Re-evaluating the Turkish Catastrophe Insurance Pool, 39 Disasters 782 (2015).

Figure 5.5 *Organizational chart of the Turkish Catastrophe Insurance Pool*

the management of international reinsurance companies, like Munich Re.[123] The claims payment of the TCIP is dependent on international reinsurance and on the amount of funds collected (partially from the government[124]). The board of directors represents the government, experts, and insurance companies. The administrative body of the TCIP is the General Directorate of Insurance within the Prime Ministry Under-Secretariat of the Treasury, but the business operation is managed

---

Failures of Compulsory Risk Mitigation: Re-evaluating the Turkish Catastrophe Insurance Pool, 39 Disasters 782 (2015).

[123] Burcak Başbuğ-Erkan and Ozlem Yilmaz, *Successes and Failures of Compulsory Risk Mitigation: Re-evaluating the Turkish Catastrophe Insurance Pool,* 39 Disasters 782 (2015).

[124] It would only be triggered by an event equivalent to an earthquake in Istanbul with a 200-year return period (technically, an earthquake with an exceedance probability of 0.5 percent). See Eugene Gurenko, Earthquake Insurance in Turkey: History of the Turkish Catastrophe Insurance Pool xi–xii (2006).

by Milli Reasürans ("operational manager"), a national reinsurance company.[125]

Loss-sensitive premiums Since the business operation of the TCIP follows a market-oriented approach, and its underwritten risks are transferred to international reinsurers, it is reasonable for international reinsurers to charge loss-sensitive premiums to control the moral hazard of the TCIP. Loss-sensitive premiums require that reinsurance premiums should reflect an actuarially fair cost, and they constrain the TCIP to underwrite appropriately. Due to the financial burden of reinsurance, the TCIP adopts a differential risk-based pricing approach and imposes construction maintenance obligations on the insured to mitigate underwritten losses.[126]

Duty of utmost good faith Primary insurers play a different role in the TCIP compared to their role in the French CCR and Japanese JERS. Primary insurers act as agents to the TCIP, and the pool assumes all the earthquake risks.[127] This raises a principal-agent problem, which is a classic situation where the duty of utmost good faith should and would apply. Furthermore, the TCIP transfers risk to international reinsurers, and international reinsurers also require the TCIP to perform the duty of utmost good faith.

Providing risk management services Reinsurers play an important role as consultants, especially in the conception of the TCIP. In fact, the TCIP was formed with the cooperation of the World Bank, the Turkish Government, Milli Re, reinsurance brokers, and Munich Re.[128] International reinsurers play an important role in providing risk management services and contribute to the operation of the TCIP and catastrophe risk management in Turkey.

---

[125] Burcak Başbuğ-Erkan and Ozlem Yilmaz, *Successes and Failures of Compulsory Risk Mitigation: Re-evaluating the Turkish Catastrophe Insurance Pool*, 39 Disasters 782 (2015). All of its business functions—from sales to reinsurance to claim management—are subcontracted to the private insurance industry, and the TCIP has no public employees. See Eugene Gurenko, Earthquake Insurance in Turkey: History of the Turkish Catastrophe Insurance Pool xi–xii (2006).

[126] Article 14 of Governmental Decree Law No. 587 on Compulsory Earthquake Insurance stipulates, "The owner who causes or allows the building and each independent section thereof to be altered contrary to the related design and in a way that will affect the load-bearing system, loses his entitlement to compensation in as much as the actual loss arises or increases because of such reason".

[127] Johann-Adrian von Lucius, *A Reinsurer's Perspective on the Turkish Catastrophe Insurance Pool (TCIP)* in Catastrophe Risk and Reinsurance: A Country Risk Management Perspective 217 (Eugene N. Gurenko ed., 2004).

[128] World Forum of Catastrophe Programmes, Natural Catastrophes Insurance Cover: A Diversity of Systems 163–164 (2008).

Table 5.1   Comparison of regulation by government-sponsored reinsurance

	French CCR	Japanese JERS	Turkish TCIP
Loss-sensitive premiums	Partially	No	Yes
The duty of utmost good faith	Yes	Yes	Probably
Providing risk management services	Not clear	Not clear	Yes
Indirect regulation of insureds	No	Yes	No

Indirect regulation of insureds Since international reinsurers, such as Munich Re, are licensed to conduct business in Turkey, there is no need for a fronting agreement arrangement. There is no empirical evidence that the TCIP indirectly regulates insureds.

D.   Concluding Remarks

Controlling moral hazard and providing incentives for loss control benefit both reinsurers and primary insurers. Such efforts will encourage ceding companies to regulate behaviors of policyholders, decrease costs for ceding companies, and enhance profits for reinsurers. It is a win-win strategy for both reinsurers and primary insurers. Compared to private reinsurers, government-sponsored reinsurance faces more challenges in loss control due to political pressures and other constraints. Table 5.1 summarizes the regulation by government-sponsored reinsurance among the three countries in the preceding discussion.

Table 5.1 shows that no government-sponsored reinsurance fully performs the ideal regulatory techniques. It appears that the Turkish TCIP is subject to less moral hazard than the French CCR and the Japanese JERS. The TCIP cedes risks to international reinsurers following a loss-sensitive-premiums approach and thus has more incentives to underwrite appropriately, such as identifying "bad risks", enforcing building codes, and educating the public to raise its awareness of catastrophe risk. Meanwhile, international reinsurers not only helped found the TCIP but also work as consultants providing risk management services. The application of regulatory techniques of reinsurance helps the TCIP work sustainably. In this way, the TCIP provides a model solution, especially for developing and middle-income countries where severe catastrophe risks exist.

Different from the TCIP, the French CCR and the Japanese JERS are both government-sponsored reinsurance institutions and are not involved with other private reinsurance companies. Although they do not adopt loss-based premiums due to political pressures, they are better at enforcing primary insurers' duty of utmost good faith than the TCIP. The CCR system is particularly suitable to France for several reasons. The first reason is cultural influence. In France, people value the national solidarity principle and are tolerant of cross-subsidies between different classes of risk and different regions, both of which guarantee a single-rate price for reinsurance. The second reason is social adequacy and affluence. As France is a developed and high-income country, the French government has more capacity to sponsor policyholders. The third reason is the moderate exposure to disasters. None of the 25 worst natural disasters recorded, including earthquakes, typhoons, and tsunamis, occurred in France.[129] In addition, during the last several decades (1970–2013), none of the natural disasters that caused the top 10 insured catastrophe losses occurred in France.[130]

Japan faces more severe catastrophe risks than France because of the frequent occurrence of earthquakes and tsunamis. The establishment of the Japanese JERS is a compromise between the government and the insurance industry: the government provides reinsurance capacity as a last resort and facilitates insurance affordability.[131] There is no doubt that the JERS follows a general fair-value principle for price setting. Under such a situation, the JERS pays more attention to monitoring primary insurers' performance of duty of utmost good faith and indirect regulation of insureds to control moral hazard and to mitigate losses.

VI.   EXPANDING REGULATION BY GOVERNMENT-SPONSORED CATASTROPHE REINSURANCE TO CHINA?

This chapter has reviewed the imperfections of private reinsurance, which are mainly due to the apparent shortage of reinsurance capital, especially during hard markets. Also discussed were government-sponsored reinsurance programs in France, Japan, and Turkey, which represent both high-income and middle-income countries. The focus now is to explore the

---

[129]   Josef, *25 Worst Natural Disasters Ever Recorded* (August 26, 2013), available at http://list25.com/25-worst-natural-disasters-recorded/5/.

[130]   Swiss Re, *Natural Catastrophes and Man-Made Disasters in 2013: Large Losses from Floods and Hail; Haiyan Hits the Philippines*, Sigma No. 1, at 5 (2014).

[131]   Yuichi Takeda, *Government as Reinsurers of Last Resort: The Japanese Experience* in Catastrophe Risk and Reinsurance: A Country Risk Management Perspective 225 (Eugene N. Gurenko ed., 2004).

possibility of expanding regulation by reinsurance to China in facilitating government intervention in catastrophe risk management.

A.   Justification of Government-sponsored Reinsurance in China

Section IV has explained the imperfections of the private reinsurance market for catastrophe risks, but these market failures are not sufficient to justify any and all government intervention: there are many different forms of government-provided reinsurance, some of which may be ineffective (no efficiency gains achieved) or even detrimental (causing efficiency losses).[132] One popular approach to government intervention is to provide a government bailout to victims, including ad hoc direct payment and establishing compensation funds. This type of ex post bailout is known as the Whole-Nation System, but it is generally seen as problematic, as was discussed in Chapter 1.[133] Another popular approach to government intervention is government-provided insurance. Compared with ex post government bailouts, this type of government intervention appears more attractive, since an ex ante insurance approach could accumulate reserves and provide incentives to mitigate losses before disasters if associated with risk-based premiums. However, this type of government intervention is also generally seen as problematic, as discussed in Chapters 2, 3, and 4.[134] Even in China, where private catastrophe insurance has not yet developed, the government should facilitate private insurance rather than provide government insurance. The Chinese government could adopt a reinsurance regime for catastrophes or provide reinsurance capacity as

---

[132]   David Cummins and Olivier Mahul, Catastrophe Risk Financing in Developing Countries: Principles for Public Intervention 76 (2009). See also W. Neil Adger, Nigel Arnell and Emma Tompkins, *Successful Adaptation to Climate Change*, 15 Global Environmental Change 85 (July 2005).

[133]   Simply speaking, the problems include undercutting potential victims' incentive for risk prevention and loss mitigation and posing a heavy fiscal burden for the government and causing negative distributional effects, leading to political inefficiencies, etc.

[134]   For example, government-provided insurance always delivers a subsidy that private insurance cannot provide, and creates two distortions: (1) regressive redistribution that favors affluent policyholders and (2) inefficient investment in residential property locating too many assets in vulnerable areas. Some scholars have reviewed and examined two government-provided insurance programs, the National Flood Insurance Program and Florida's state-owned Citizens Insurance, and found that both perceptions of government-provided insurance performance along two normative metrics, fairness and efficiency, are wrong. See Omri Ben-Shahar and Kyle Logue, *The Perverse Effects of Subsidized Weather Insurance*, 68 Stanford Law Review 571 (2016).

a last resort. Such arrangements and intervention provide considerable incentive for primary insurers to control moral hazard and mitigate losses associated with catastrophic disasters.

China has now begun to stimulate the development of catastrophe insurance to complement government action in addressing catastrophe risks. Government provision of reinsurance capacity would also help stimulate the supply of and demand for private insurers and reinsurers.

In 2013, the Third Plenary Session of the Eighteenth Communist Party of China Central Committee promulgated the "Decision of the Central Committee of the Communist Party of China on Some Major Issues concerning Comprehensively Deepening the Reform", which expressly stated that "we will establish an insurance system for catastrophe risks". However, the current insurance industry has few incentives to underwrite catastrophe risks partly due to scarce insurance and reinsurance capacity. In 2014, catastrophe insurance program trials were launched in Shenzhen, in the Pearl River Delta (a densely populated metropolitan area and also one of the world's most disaster-prone regions), and in the Chuxiong region in the southwestern province of Yunnan, known to be prone to earthquakes.[135] All of these programs are initiated and sponsored by the government rather than private insurers.

Private catastrophe insurance remains one of the least developed areas of insurance in China. For example, after the 2008 Great Sichuan Earthquake, only 0.3 percent of the total losses were covered by insurance companies.[136] Private insurers do not have the capital to fully cover catastrophe losses. The total net capital of China's property insurance companies is much lower than the total amount of losses caused by natural disasters. Table 5.2 shows the extent of this shortfall. Moreover, as of 2015, the China Insurance Regulatory Commission has implemented China's Risk-Oriented Solvency System.[137] The new solvency regime, as under the EU Solvency II Directive, requires insurers to hold sufficient

---

[135]   *China Says Testing Catastrophe Insurance System*, Reuters (August 20, 2014), available at http://www.businessinsurance.com/article/20140820/NEWS04/140829990?AllowView=VDl3UXk1T3hDUFNCbkJiYkY1TDJaRUt0ajBRV0ErOVVHUT09#.

[136]   *Establishing Catastrophe Insurance System faces acceleration*, China Youth Daily (14 March 2011), available at http://zqb.cyol.com/html/2011-03/14/nw.D110000zgqnb_20110314_1-05.htm?div=-1.

[137]   Wenhui Chen, *C-ROSS under the Market-oriented Reform and Economy Globalization*, Risk Dialogue Magazine (2014), available at http://media.swissre.com/documents/CROSS_under_the_market_ChenWenhui_Dec15.pdf.

Table 5.2   Net capital of main Chinese property insurers compared to natural disaster losses (US$ billions)

	2007	2008	2009	2010
Net capital of main insurers	5.5	5.1	6.9	9.0
Natural disaster losses	38.1	189.5	40.1	86.1

Source: Yearbook of China Insurance (2008–2011).

asset capital in their reserves, especially capital for catastrophe risks.[138] In order to underwrite catastrophe risks, insurers increasingly demand more financial capacity and that a significant portion of the insured losses are met by reinsurers, either through private reinsurance or government-sponsored reinsurance.

However, private reinsurance currently does not provide strong support for catastrophe insurance in China. At present, the China Reinsurance (Group) Corporation (its predecessor, the People's Insurance Company of China Reinsurance, was created in 1996) is the only domestic reinsurer in China, with consolidated total assets of around $30 billion and net assets of $8.6 billion.[139] Its net capital is much lower than the annual losses caused by natural disasters. As for international reinsurers, although China's reinsurance market became open to foreign reinsurance companies after China's entry into the World Trade Organization, only a few reinsurance companies, such as Swiss Re and Munich Re, have established business operations in China. By 2013 there were only eight foreign reinsurers that had registered branches in China,[140] and they are only in the initial stages of reinsuring risks. They might have few incentives to cover sophisticated and complicated risks of domestic insurers, and thus it is difficult for catastrophe insurers to cede their risks to international reinsurers in such situations.

When private insurers and reinsurers are encouraged or even required to cover catastrophe risks, they will demand government intervention, which

---

[138]   Wenhui Chen, *C-ROSS under the Market-oriented Reform and Economy Globalization*, Risk Dialogue Magazine (2014), available at http://media.swissre.com/documents/CROSS_under_the_market_ChenWenhui_Dec15.pdf.

[139]   China Re, Annual Report (2014), available at http://www.chinare.com.cn/zhzjt/resource/cms/2015/08/2015082709085075513.pdf.

[140]   CPCR, Overview of Chinese Reinsurance Market (2013), available at https://www.casact.org/education/spring/2013/handouts%5CPaper_1680_handout_962_0.pdf.

could provide the government sponsorship with deep credit capacity. Government-sponsored reinsurance, which works as the last resort for private insurers and reinsurers, could be a reasonable choice since it could expand their capacity and further stimulate the growth of private insurers and reinsurers.

B.   Effectiveness of Regulation by Catastrophe Reinsurance

The Chinese government could provide reinsurance capacity as a last resort to catastrophe risk management due to its deep credit capacity, as was discussed in Chapter 1.[141] What is less clear is how to achieve the proper regulatory goals, as discussed in sections III and V above. In 2014, China launched its first catastrophe insurance pilot in Shenzhen ("the Shenzhen Model"). Therefore, the possibility and feasibility of regulation by reinsurance in China will be explored through the examination of its regulatory techniques in the Shenzhen Model.

Shenzhen was selected for the pilot because its geographical position makes it prone to major catastrophe threats[142] and it has a large number of valuable assets.[143] The catastrophe insurance framework of the Shenzhen Model includes three tiers: the first tier is the government catastrophe insurance assistance, which is bought by the Shenzhen municipal government from private insurers, with the beneficiaries being all residents of Shenzhen City; the second tier is a catastrophe fund mainly sponsored by the Shenzhen government and social donations; and the third layer is commercial catastrophe insurance.[144]

---

[141]   For example, the Whole-Nation System is the performance of its credit capacity.

[142]   Frequently occurring disasters in Shenzhen include but are not limited to: heavy winds (gales, strong gales, and fresh gales), rainstorms, lightning strikes, floods, waterlogging, tornados, typhoons, tsunamis, hail, landslides, mudslides, cliff falls, land subsidence, squall lines, and earthquakes of more than 4.5 magnitude. See ICC, Catastrophe Insurance Framework of Shenzhen City (2015), available at http://wenku.baidu.com/link?url=oLT1RmQ3BXgfW49ETc-Drhv6S1pOb8dOA5E3Y-OVZgCAkJrTD-aiBaF1doiXOq9Xsb1rLoty4IP-b1dPBKzZY2eiNgZex52Gfzpd-heyzEIt3.

[143]   Shenzhen is a megacity with approximately 15 million residents. It is China's first and one of the most successful Special Economic Zones and had a GDP totaling $260.48 billion in 2014. See Yisha Hou, *Promoting the Construction of Shenzhen Catastrophe Insurance System*, Disaster Reduction in China 42 (2015).

[144]   *China says testing catastrophe insurance system*, Reuters (August 20, 2014), available at http://www.businessinsurance.com/article/20140820/NEWS04/14082 9990?AllowView=VDl3UXk1T3hDUFNCbkJiYkY1TDJaRUt0ajBRV0ErOVV HUT09#. See also http://xw.sinoins.com/2014-07/10/content_120490.htm.

The first two tiers of the Shenzhen Model represent a hybrid private/ municipal arrangement.[145] The first tier is the government catastrophe assistance insurance, which is bought by the Shenzhen municipal government to supply basic assistance for all residents.[146] The Shenzhen government purchases its catastrophe insurance products from People's Insurance Company of China (PICC) Shenzhen branch, whose coverage is underwritten by Swiss Re, China Re, and Taiping Re.[147] This insurance functions as indemnity insurance: the insurance company will pay for actual medical costs and pension costs due to victims' injuries or deaths caused by the disasters. Payments are only available for bodily injury and death but not property damage.[148] Payment for a single disaster event is capped at 2 billion yuan (RMB) with maximum individual claims of 100,000, approximately US$15,710.[149] The pilot program clearly does not contemplate a claim being made by a large number of citizens for any single event. Since coverage for a single disaster event is capped at RMB 2 billion, the program will be able to deliver maximum individual relief to 25,000 individuals in the event of a disaster.[150]

The second tier of the Shenzen Model is a catastrophe fund managed

---

[145]   The details of the Shenzhen Model are mainly drawn from one of the author's co-authored papers: Anastasia Telesetsky and Qihao He, *Climate Change Insurance and Disasters: Is the Shenzhen Parametric Social Insurance a Model for Adaptation?*, 43 Boston College Environmental Affairs Law Review 485 (2016).

[146]   *Catastrophe Insurance Framework of Shenzhen City*, PICC (October 12, 2015), available at http://wenku.baidu.com/link?url=oLT1RmQ3BXgfW49ETc-Drhv6S1pOb8dOA5E3YOVZgCAkJrTD-aiBaF1doiXOq9Xsb1rLoty4IP-b1dPB KzZY2eiNgZex52GfzpdheyzEIt3 [https://perma.cc/RPG6-B9Z5].

[147]   *China Re, Swiss Re, and Taiping Re Underwrite the Shenzhen Catastrophe Reinsurance Policies* (July 6, 2014), available at http://xw.sinoins.com/2014-07/16/ content_121575.htm (underwriting from multiple reinsurance agencies).

[148]   Martin Li and Yin Ran, *SZ launches 1st disaster insurance*, Shenzhen Daily (July 17, 2014), available at http://szdaily.sznews.com/html/2014-07/10/con-tent_2936724.htm.

[149]   *Catastrophe Insurance Framework of Shenzhen City*, PICC (October 12, 2015), available at http://wenku.baidu.com/link?url=oLT1RmQ3BXgfW49ETc-Drhv6S1pOb8dOA5E3YOVZgCAkJrTD-aiBaF1doiXOq9Xsb1rLoty4IP-b1dPBKzZY2eiNgZex52GfzpdheyzEIt3 [https://perma.cc/RPG6-B9Z5]; Martin Li and Yin Ran, *SZ Launches 1st Disaster Insurance*, Shenzhen Daily (July 17, 2014), available at http://szdaily.sznews.com/html/2014-07/10/content_2936724. htm [https://perma.cc/X4YQ-PSP3].

[150]   See *Catastrophe Insurance Framework of Shenzhen City*, PICC (October 12, 2015), available at http://wenku.baidu.com/link?url=oLT1RmQ3BXgfW49ETc-Drhv6S1pOb8dOA5E3YOVZgCAkJrTD-aiBaF1doiXOq9Xsb1rLoty4IP-b1dPBKzZY2eiNgZex52GfzpdheyzEIt3 [https://perma.cc/RPG6-B9Z5].

by the city.[151] Capital sources of the catastrophe fund include government appropriation, social donation, and investment earnings.[152] Where catastrophe losses exceed the cap of RMB 2.5 billion available in the first tier of insurance, the catastrophe fund should provide extra coverage.[153] In addition, Shenzhen City intends to issue catastrophe insurance-linked securities, such as catastrophe bonds, and thus transfer catastrophe risk to capital markets.[154] The government-funded policy covering the first and second tiers of the insurance framework is limited in scope and covers 15 events including heavy winds, rainstorms, lightning strikes, floods, waterlogging, tornadoes, typhoons, tsunamis, hail, landslides, mudslides, cliff slides, land subsidence, severe thunderstorms, and earthquakes of more than 4.5 magnitude.[155]

The final tier of the Shenzhen Model is private catastrophe insurance to cover property losses.[156] The third layer has not yet been finalized while the Shenzhen municipal government encourages private insurers and reinsurers to cover the losses that exceed the cap available in the first and second tiers. The PingAn Insurance Company, whose headquarters are in Shenzhen, is starting to design and sell relative catastrophe insurance products to the residents of Shenzhen.[157] Commercial primary insurers,

---

[151]   *Catastrophe Insurance Framework of Shenzhen City*, PICC (October 12, 2015), available at http://wenku.baidu.com/link?url=oLT1RmQ3BXgfW49ETc-Drhv6S1pOb8dOA5E3YOVZgCAkJrTD-aiBaF1doiXOq9Xsb1rLoty4IP-b1dPB-KzZY2eiNgZex52GfzpdheyzEIt3 [https://perma.cc/RPG6-B9Z5]. See also Martin Li and Yin Ran, *SZ Launches 1st Disaster Insurance*, Shenzhen Daily (July 17, 2014), available at http://szdaily.sznews.com/html/2014-07/10/content_2936724. htm [https://perma.cc/X4YQ-PSP3] (describing a RMB 30-million catastrophe fund).

[152]   See Martin Li and Yin Ran, *SZ Launches 1st Disaster Insurance*, Shenzhen Daily (July 17, 2014), available at http://szdaily.sznews.com/html/2014-07/10/content_2936724.htm [https://perma.cc/X4YQ-PSP3].

[153]   *Catastrophe Insurance Framework of Shenzhen City*, PICC (October 12, 2015), available at http://wenku.baidu.com/link?url=oLT1RmQ3BXgfW49ETc-Drhv6S1pOb8dOA5E3YOVZgCAkJrTD-aiBaF1doiXOq9Xsb1rLoty4IP-b1dPBKzZY2eiNgZex52GfzpdheyzEIt3 [https://perma.cc/RPG6-B9Z5].

[154]   Yisha Hou, *Promoting the Construction of Shenzhen Catastrophe Insurance System*, Disaster Reduction in China 42 (2015).

[155]   Yisha Hou, *Promoting the Construction of Shenzhen Catastrophe Insurance System*, Disaster Reduction in China 42 (2015).

[156]   *Catastrophe Insurance Framework of Shenzhen City*, PICC (October 12, 2015), available at http://wenku.baidu.com/link?url=oLT1RmQ3BXgfW49ETc-Drhv6S1pOb8dOA5E3YOVZgCAkJrTD-aiBaF1doiXOq9Xsb1rLoty4IP-b1dPBKzZY2eiNgZex52GfzpdheyzEIt3 [https://perma.cc/RPG6-B9Z5].

[157]   Yisha Hou, *Promoting the Construction of Shenzhen Catastrophe Insurance System*, Disaster Reduction in China 42 (2015).

like PingAn Insurance Company, have strong incentives to transfer catastrophe excess risks to reinsurers. However, reinsurers like Swiss Re, China Re, and Taiping Re, which are involved in the Shenzhen Model, offer technical assistance and reinsurance only for the first tier.[158] Therefore, government-sponsored reinsurance could be regarded as last resort for the private primary insurers. If government-sponsored reinsurance is adopted in the future, it is important that regulatory techniques are applied to control moral hazard of primary insurers and to mitigate losses.

Loss-sensitive premiums In the first tier of the Shenzhen Model, the government buys insurance products from insurance companies (the PICC, Shenzhen branch) rather than acting as a reinsurer. The PICC then cedes a large portion of its underwriting of catastrophe risks to Swiss Re, China Re, and Taiping Re, according to the quota share treaties.[159] These reinsurers charge the PICC loss-sensitive premiums. Under these premiums, the PICC has incentives to control moral hazard and mitigate losses. For example, the PICC uses 5 percent of the premium it charges the government to fund disaster research, disaster prevention, disaster emergency relief drills, and disaster emergency advertising; submits to the government a quarterly report of current disaster and claims payments and an annual report of disaster risk management; offers advice on risk prevention, emergency management, and disaster relief to the municipal government; and establishes and operates a disaster data base for disaster analysis and prevention.[160] In tandem with experts, insureds, and other stakeholders, the PICC has identified the technical and economic parameters of catastrophe risks and has developed system-wide technologies of loss prevention.

Furthermore, loss-sensitive premiums also encourage primary insurers to regulate policyholders' behavior in relation to loss mitigation. The PICC offers the Shenzhen government a discounted premium for taking cost-effective mitigation measures. For example, the PICC provides that if the annual loss ratio (actual payment amount/total premium) is less than 10 percent, then the premium the following year will be discounted by 10

---

[158]   *Swiss Re Works with Government Bodies in Mitigating Natural Catastrophe Risks in China*, Swiss Re (August 13, 2014), available at http://www.swissre.com/china/Swiss_Re_works_with_government_bodies_in_mitigating_natural_catastrophe_risks_in_China.html [https://perma.cc/88RV-NLHR].

[159]   *China Re, Swiss Re, and Taiping Re Underwrite the Shenzhen Catastrophe Reinsurance Policies* (July 6, 2014), available at http://xw.sinoins.com/2014-07/16/content_121575.htm.

[160]   *Catastrophe Insurance Framework of Shenzhen City*, PICC (October 12, 2015), available at http://wenku.baidu.com/link?url=oLT1RmQ3BXgfW49ETc-Drhv6S1pOb8dOA5E3YOVZgCAkJrTD-aiBaF1doiXOq9Xsb1rLoty4IP-b1dPBKzZY2eiNgZex52GfzpdheyzEIt3 [https://perma.cc/RPG6-B9Z5].

percent; if the loss ratio is less than 10 percent in two consecutive years, the third year's premium will be discounted by 20 percent; if the loss ratio is less than 10 percent in three consecutive years, the fourth year's premium will be discounted by 30 percent.[161] Although a 30 percent discount may raise solvency concerns, there is no data on whether this pricing structure is sustainable.

Although the third tier of the Shenzhen Model has not been finalized, its design and goals are to encourage and attract private insurers to underwrite losses that exceed the cap of the first two tiers. According to the comparative discussion on the French CCR, the Japanese JERS, and the Turkish TCIP, it is advisable not to emulate their pricing practices in the third layer of the Shenzhen Model. If the form of government-sponsored reinsurance follows the model of the French CCR or the Japanese JERS to sponsor primary insurers, it will not be easy to charge loss-sensitive premiums to control moral hazard since political pressure or other reasons would not prevent it from repeating their mistakes in subsidizing premiums. An alternative choice is the Turkish TCIP model. Private insurers could work in the same way as the TCIP and cede risks to international reinsures following a loss-sensitive-premiums approach while the government would provide contingent liquidity support when the payments of claims exceeded insurers' capacity.

Duty of utmost good faith The quota share treaty between insurers and reinsurers encourages the PICC's performance of the duty of utmost good faith, since the PICC has to retain some portion of the risks itself. In contrast, the typical long-term relationship mechanism between insurers and reinsurers, which is closely associated with utmost good faith, may not be workable in the Shenzhen Model. The current Shenzhen Model is a temporary trial project and lacks authorizing legislation.[162] Without explicit legislative provisions, the prospect of the Shenzhen Model is uncertain. The Shenzhen municipal government may cease to buy catastrophe insurance policies in future years. If the government does not buy insurance, there will be no opportunity for a long-term relationship between the PICC and rein-surers and thus no basis for operation of the duty of utmost of good faith.

---

[161]   *Catastrophe Insurance Framework of Shenzhen City*, PICC (October 12, 2015), available at http://wenku.baidu.com/link?url=oLT1RmQ3BXgfW49ETc-Drhv6S1pOb8dOA5E3YOVZgCAkJrTD-aiBaF1doiXOq9Xsb1rLoty4IP-b1dPBKzZY2eiNgZex52GfzpdheyzEIt3 [https://perma.cc/RPG6-B9Z5].
[162]   Xing Shi, *Inspirations of Shenzhen Catastrophe Insurance Pilot*, 21st Century Economic Report (2014), available at http://insurance.hexun.com/2014-10-11/169210867.html.

Therefore, in order to maintain the long-term relationship between primary insurers and reinsurers, it may be better to enact authorizing legislation.

Providing risk management services Like the Turkish TCIP, reinsurers, especially international reinsurers like Swiss Re and Munich Re, play an important role as consultants in providing risk management services under the Shenzhen Model. For example, Swiss Re initiated a Parametric Insurance Solutions for Disaster Relief System Reform research program in 2013 as a sponsor for the China Development and Research Foundation.[163] This research program has helped Swiss Re become a technical advisor and a leading reinsurer for the Shenzhen Model.[164] It is worth emulating the cooperation with international reinsurers, which can provide risk management services for further pilot programs in China.

VII. CONCLUSION

Government-sponsored reinsurance can address failures in the private market for catastrophe insurance due to the deep credit capacity of the government. Considered the corollary of the regulation-by-insurance idea,[165] as the title of this chapter suggests, government-sponsored reinsurance can also regulate primary insurers' behavior in risk mitigation and risk management through reinsurers' regulatory techniques.

Currently, affected parties of natural disasters, especially pilot catastrophe insurers, are demanding government sponsorship of their catastrophe losses in China. Considering the reform of the Whole-Nation System, there is a pressing need for the Chinese government to provide reinsurance capacity as the new government-intervention approach. Moreover, regardless of which type of government intervention the Chinese government adopts, it is necessary to exert the role of reinsurance in regulating primary insurers through reinsurance regulatory techniques.

---

[163] Swiss Re, *Swiss Re Works with Government Bodies in Mitigating Natural Catastrophe Risks in China* (August 13, 2014), available at http://www.swissre. com/china/Swiss_Re_works_with_government_bodies_in_mitigating_natural_cat astrophe_risks_in_China.html.

[164] Ailin Liu, *Risks of the First Catastrophe Insurance Policy are Ceded to China Re, Swiss Re and CPIC Re* (2014), available at http://finance.sina.com.cn/money/insurance/bxdt/20140716/025719714880.shtml.

[165] Aviva Abramovsky, *Reinsurance: The Silent Regulator?*, 15 Connecticut Insurance Law Journal 345 (2009) ("Just as insurance is often viewed as having a regulatory effect on insured industries, so too should reinsurance be considered as having a regulatory effect on its reinsureds").

6. Innovations in insurance markets and securitization of catastrophe risk: experiences and lessons to learn[1]

I. INTRODUCTION

After years of deliberation, especially following the 2008 Great Sichuan Earthquake, China accelerated the promotion of a catastrophe insurance system. In 2012, one of the main topics in the Fourth National Finance Working Conference was establishing such a system.[2] In 2013, the Third Plenary Session of the Eighteenth Communist Party of China Central Committee promulgated the "Decision of the Central Committee of the Communist Party of China on Some Major Issues concerning Comprehensively Deepening the Reform", which expressly stated that "an insurance system for catastrophe risks should be established".[3] In 2014, catastrophe insurance program trials were launched in Shenzhen, the Pearl River Delta (a densely populated metropolitan area and also one of the world's most disaster-prone regions), and the Chuxiong

---

[1] The previous version of this chapter was published with co-author Ruohong Chen in 8 Frontiers of Laws in China 523 (2013).

[2] The National Finance Working Conference is the supreme finance conference in China, and it has decided important finance issues, such as establishing the China Securities Regulation Commission (CSRC) and the China Insurance Regulation Commission (CIRC) at the First Conference; establishing the China Banking Regulation Commission (CBRC) and selecting three major state-owned commercial banks (China Banks, China Construction Bank, and China Commerce Bank) to begin initial public offerings (IPOs), at the Second Conference; and establishing the China Investment Corporation, at the Third Conference. The conference is held approximately every five years in Beijing. See http://finance. ce.cn/sub/2011/jrgzhy/.

[3] Johanna Hjalmarsson and Mateusz Bek, *Legislative and Regulatory Methodology and Approach: Developing Catastrophe Insurance in China* in Insurance Law in China 191 (Johanna Hjalmarsson and Dingjing Huang eds., 2015).

Region in the southwestern province of Yunnan, known to be prone to earthquakes.[4]

However, catastrophic perils challenge insurers: an increasing number and severity of weather-related extreme events combine with considerable uncertainty as to where and how those losses will develop.[5] Many insurers may be reluctant to underwrite catastrophic exposures and have even cut back coverage, for example after Hurricane Andrew in the United States.[6] There may be ways of lowering the cost of providing catastrophe insurance that will make insurers more willing to supply it. Alternative Risk Transfers (ARTs) are an example of such cost-lowering innovations.[7] Securitization of catastrophe risk links the insurance industry to the capital markets, and offers the potential to expand the capacity of the insurance industry to take on catastrophe risks. Securitization of catastrophe risks has been undertaken in the United States since the mid-1990s, particularly in the form of catastrophe bonds.

In 2015, China's first company issued its first catastrophe bond—Panda Re Ltd (Series 2015-1)—to cover natural disaster risks by the state-owned reinsurer China Reinsurance Company.[8] This bond was issued on the international market, not in the domestic market, through a special

---

[4] *China Says Testing Catastrophe Insurance System*, Reuters (2014), available at http://www.businessinsurance.com/article/20140820/NEWS04/140829990?AllowView=VDl3UXk1T3hDUFNCbkJiYkY1TDJaRUt0ajBRV0ErOVVHUT09#.

[5] Joseph MacDougald and Peter Kochenburger, *Insurance and Climate Change*, 47 John Marshall Law Review 719 (2013).

[6] Joseph MacDougald and Peter Kochenburger, *Insurance and Climate Change*, 47 John Marshall Law Review 719 (2013).

[7] Alternative Risk Transfer (ART) is the use of risk management tools other than traditional insurance or reinsurance to provide risk-bearing entities with risk protection. The key difference between ART and the traditional insurance market-place is that insurance and reinsurance markets provide catastrophic risk coverage whereas ARTs provide additional financial capacity for insurance coverage. See E. Michel-Kerjan and F. Morlaye, *Extreme Events, Global Warming, and Insurance-linked Securities: How to Trigger the "Tipping Point"*, 33 The Geneva Papers on Risk and Insurance—Issues and Practice 153 (2008); E. Banks, Alternative Risk Transfer: Integrated Risk Management through Insurance, Reinsurance, and the Capital Markets 49 (2004). Christopher Kampa, Alternative Risk Transfer: The Convergence of the Insurance and Capital Markets, Part I, A Broad Overview (2010). Robert J. Rhee, *Terrorism Risk in a Post-9/11 Economy: The Convergence of Capital Markets, Insurance, and Government Action*, 37 Arizona State Law Journal 435 (2005).

[8] Matthew Lerner, *Panda Re Catastrophe Bond Covers China risks* (2015), available at http://www.businessinsurance.com/article/20150702/NEWS06/150709961?tags=58|338|158|166|93|143|83|76|70|64.

purpose vehicle set up in Bermuda.[9] This is China's first step toward linking catastrophe insurance to capital markets and embracing securitization of catastrophe risk.

The objective of this chapter is to provide an overview of securitization of catastrophe risk and analyse the supply and demand dynamics of financial innovations in the market for catastrophe insurance risk transfer, and further, to describe the international experience from which China could learn and the domestic obstacles China should conquer. The chapter will first introduce the general practices of securitization of catastrophe risk, exemplified by catastrophe bonds, since they have been the predominant form of insurance-linked securities until now. Next, the supply and demand dynamics of insurance-linked securities will be discussed. Finally, the chapter concludes with an analysis of the legal obstacles and financial market shortcomings when issuing insurance-linked securities in China's domestic market, and some proposed solutions.

II.   INSURANCE SECURITIZATION BASICS

A.   Insurance Securitization

In its simplest form, securitization is "the process of pooling assets for purposes of investment on behalf of individual investors looking to purchase a fractional share of the pool".[10] In other words, from the perspective of the entity issuing the security, securitization is the process of removing assets, liabilities, or cash flow from the entity's balance sheet and conveying them to third parties through tradable securities.[11] Thus, securitization of insurance risk is the process of transferring insurance risks from insurers and conveying them to third parties through tradable securities. The idea of securitizing insurance risk was first suggested by Robert Goshay and Richard Sandor in the 1970s.[12] In the 1990s, investment banks started to securitize insurance risk in practice and brought out products known

---

[9]   Xinhua News Agency, China issues catastrophe bond overseas for first time (2015), available at http://news.xinhuanet.com/english/2015-07/02/c_134376924. htm.

[10]   Todd V. McMillan, *Securitization and Catastrophe Bonds: A Transactional Integration of Industries Through a Capacity-Enhancing Product of Risk Management*, 8 Connecticut Insurance Law Journal 131 (2002).

[11]   Erik Banks, Catastrophic Risk, Analysis and Management 111 (2005).

[12]   See Robert Goshay and Richard Sandor, *An Inquiry into the Feasibility of a Reinsurance Futures Market*, 5 Journal of Business Finance 56 (1973).

as insurance-linked securities. These securities, which have also been described as the securitization of catastrophe risk,[13] are a mainstay of the ART market.

Traditionally, insurance companies hold capital in reserve or cede to reinsurance companies to manage risk. Insurance-linked securities enable insurers to transfer risks to the capital market and provide insurers additional capacity for coverage.[14] Meanwhile, insurance-linked securities also provide capital market investors with new investment opportunities that are traded on exchanges and over-the-counter markets.

The issuance of insurance-linked securities—catastrophe bonds, for example[15]—involves the creation of a special purpose vehicle (SPV) or a single purpose reinsurer (SPR). In standard catastrophe bond issuances, an SPV or an SPR is sponsored by an insurance company or a reinsurance company. The insurer/reinsurer pays a premium for coverage to the SPV. Meanwhile, investors purchase bonds from the SPV. Investors' principal and return are connected to a trigger catastrophe. Within the risk period, if the trigger catastrophe does not occur, investors will receive return of their principal and some form of guaranteed interest on the bonds; otherwise, in the event that a catastrophe occurs, the investors will indemnify the insurer/reinsurer from the principal to cover the loss insured against.[16] Both the premium paid by the insurer and the purchase proceeds from investors are placed in a trust account owned by the SPV and invested in accordance with the trust agreement. The trust account is designed to hold the principal invested as funding for a payout in the event that the insurer's losses pierce the layer covered by the SPV because the insurer

---

[13]   The terminology "insurance-linked securities" is interchangeable with the terminology "securitization of catastrophe risk" used in this chapter.

[14]   David Cummins, *Convergence in Wholesale Financial Services: Reinsurance and Investment Banking*, 30 The Geneva Papers on Risk and Insurance—Issues and Practice 187 (2005).

[15]   Catastrophe bonds are also called "cat bonds" or "Act of God bonds", and are the most prominent and popular form of insurance-linked securities today. See Pauline Barrieu and Luca Albertini, The Handbook of Insurance-linked Securities 9 (2009). See also US Government Accountability Office, Catastrophe Insurance Risk: The Role of Risk-linked Securities and Factors Affecting Their Use, GAO-02-941 (2002) ("Currently most risk-linked securities are catastrophe bonds"). Sylvie Bouriaux and Richard MacMinn, *Securitization of Catastrophe Risk: New Developments in Insurance-linked Securities and Derivatives* (2012), available at http://www.macminn.org/papers/Securitization%20of%20Catastrophe%20Risk. pdf ("CAT bonds remain the predominant form of catastrophe risk securitization").

[16]   Todd V. McMillan, *Securitization and Catastrophe Bonds: A Transactional Integration of Industries Through a Capacity-Enhancing Product of Risk Management*, 8 Connecticut Insurance Law Journal 131 (2002).

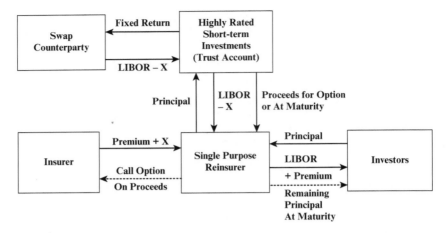

Note: LIBOR is the abbreviation for "London interbank offered rate". See David Cummins, *CAT Bonds and Other Risk-Linked Securities: State of the Market and Recent Developments*, 11 Risk Management and Insurance Review 23 (2008).

Figure 6.1   Illustration of the structure of catastrophe bonds

pays a negotiated premium amount for the coverage.[17] Furthermore, to ensure that the necessary interest rates are paid, the trustee will often enter into a swap agreement in which the swap counterparty will provide protection to the trust if the interest generated falls below the amount necessary to pay the required interest payments to the various tranche investors.[18] Figure 6.1 illustrates the expanded structure of catastrophe bonds.

The first issuance of insurance-linked securities dates back to 1995, when catastrophe options were introduced by the Chicago Board of Trade.[19] The issuance of and trading in insurance-linked securities has

---

[17]   James S. Gkonos and Braden A. Borger, *At the Crossroads of Insurance and the Capital Markets: An Analysis of the Current Regulatory Environment and New Developments in Alternative Risk Transfers*, New Appleman on Insurance: Current Critical Issues in Insurance Law (2010).

[18]   James S. Gkonos and Braden A. Borger, *At the Crossroads of Insurance and the Capital Markets: An Analysis of the Current Regulatory Environment and New Developments in Alternative Risk Transfers*, New Appleman on Insurance: Current Critical Issues in Insurance Law (2010).

[19]   US Government Accountability Office, Catastrophe Insurance Risk: The Role of Risk-Linked Securities and Factors Affecting Their Use, GAO-02-941 (2002).

been growing at a fast rate since then. By the time of the financial crisis in 2008, the total notional value of tradable insurance risk had reached $50 billion worldwide and had been growing at a rate of 40 to 50 percent a year since 1997.[20] During the crisis, the insurance-linked securities market was adversely affected, as were most other markets. However, insurance-linked securities proved to be more resilient than many other markets.[21] For example, $4.6 billion in catastrophe bonds were issued in 2010, rebounding from $5.9 billion in 2008.[22]

B.   Products of Insurance Securitization

Insurance-linked securities are considered alternative risk transfer mechanisms because they blur the boundaries of conventional insurance, reinsurance, and capital markets.[23] Generally speaking, insurance-linked securities are divided into three main types: catastrophe derivatives, contingent capital, and catastrophe bonds. Catastrophe derivatives are financial contracts and have been developed to facilitate the transfer of catastrophe risk among capital market investors.[24] Catastrophe derivatives are classified into exchange-traded derivatives and "over the counter" (OTC) derivatives contracts.[25] Contingent capital is a type of financing that is arranged before a loss occurs. When a catastrophic event (the trigger) occurs, the financier provides the insurer with capital; when

---

[20]   World Economic Forum, A World Economic Forum Report, Convergence of Insurance and Capital Markets (2008).

[21]   Aon Benfield Securities, *Insurance-Linked Securities: Capital Revolution— ILS Market Expands to New Heights* (2013), available at http://thoughtleadership. aonbenfield.com/Documents/20130830_ab_ils_annual_report_2013.pdf.

[22]   Aon Benfield Securities, *Insurance-Linked Securities: Capital Revolution – ILS Market Expands to New Heights* (2013), available at http://thoughtleadership. aonbenfield.com/Documents/20130830_ab_ils_annual_report_2013.pdf.

[23]   E. Banks, Alternative Risk Transfer: Integrated Risk Management through Insurance, Reinsurance, and the Capital Markets 58 (2004).

[24]   Partner Re, *A Balanced Discussion on Insurance Linked-Securities* (2008), available at http://www.parterre.com.

[25]   Exchange-traded derivatives are standardized derivative contracts, traded through an authorized exchange, with the exchange or its clearinghouse acting as intermediary on every contract. "Futures", "options", and "future options" constitute the primary types of exchange-traded derivatives. OTC derivatives, on the other hand, are bespoke derivative contracts that are traded directly and informally between two parties rather than via a formal exchange or other intermediary. Popular OTC derivatives include "swaps", "forwards", and "credit derivatives". See Véronique Bruggeman, Compensating Catastrophe Victims: A Comparative Law and Economics Approach 162 (2010).

no catastrophic event occurs, the insurer has no need for additional capital and the facility remains unused.[26] Catastrophe bonds are risk-linked securities that transfer catastrophe risks from insurers to investors through fully collateralized SPVs.[27] Catastrophe bonds will be discussed further below, including the issue of the introduction of catastrophe bonds into China.

The first catastrophe bond was issued in 1997 by the United Services Automobile Association (USAA); it was a hurricane bond and was designed as an alternative to reduce risk and to broaden the capacity of the insurance industry because of a disproportionate loss of $620 million during Hurricane Andrew.[28] This pioneering launch convinced regulators that investors would purchase catastrophe bonds, and regulators ultimately agreed to give market treatment to catastrophe bonds. Since then, the market for catastrophe bonds has experienced steady growth. Just as the catastrophe bonds market reached a peak in 2007, the global financial crisis spread across the world and hit the catastrophe bond market as well. In the second half of 2008, many planned catastrophe bonds transactions were postponed.[29] In 2009, the issuance of catastrophe bonds reduced further so that only $1.78 billion worth of catastrophe bonds were issued.[30] However, the issuance of catastrophe bonds has recovered since the financial crisis. As of June 30, 2013, the annual issuance volume of catastrophe bonds reached $6.7 billion, and total catastrophe bonds outstanding were at an all-time high of $17.5 billion, an increase of $2.6 billion from the previous year and surpassing the previous record of $16.2 billion on June 30, 2008[31] (Figure 6.2 and Figure 6.3).

The returns of catastrophe bondholders are connected to a trigger. A trigger may be a single event or multiple events. These events determine

---

[26]   Véronique Bruggeman, Compensating Catastrophe Victims: A Comparative Law and Economics Approach 160 (2010).

[27]   Partner Re, *A Balanced Discussion on Insurance Linked-Securities* (2008), available at http://www.parterre.com.

[28]   Erik Banks, Catastrophic Risk, Analysis and Management 112–113 (2005).

[29]   Second Conference Organized Under the Auspices of the OECD International Network on Financial Management on Large-Scale Catastrophes, *Catastrophe-Linked Securities and Capital Markets*, Organization for Economic Co-Operation and Development (2009).

[30]   Christopher Kampa, *Alternative Risk Transfer: The Convergence of the Insurance and Capital Markets*, Insurance Studies Institute (2010).

[31]   Aon Benfield Securities, *Insurance-Linked Securities: Capital Revolution— ILS Market Expands to New Heights* (2013), available at http://thoughtleadership. aonbenfield.com/Documents/20130830_ab_ils_annual_report_2013.pdf.

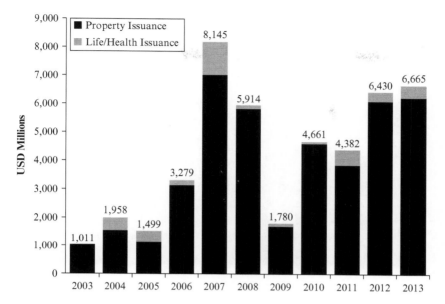

Source: Aon Benfield Securities, Insurance-Linked Securities: Capital Revolution—ILS Market Expands to New Heights (2013), available at http://thoughtleadership.aonbenfield.com/Documents/20130830_ab_ils_annual_report_2013.pdf.

Figure 6.2 Catastrophe bond issuance by year (years ending June 30)

whether the issuer can suspend the principal or interest payments either temporarily or permanently.[32] When a trigger event occurs, such as a hurricane, the principal capital is used to pay the loss. If a trigger event does not occur, the investors can earn relatively high interest from the SPV.[33] Triggers are usually structured as indemnity, index, and parametric triggers. An indemnity trigger is based on an insurer's actual losses from a particular catastrophe.[34] An index trigger is based on existing industry loss indices, which normally include numbers of risks, value by type,

[32] Christopher Kampa, Alternative Risk Transfer: The Convergence of the Insurance and Capital Markets, Part II, Non-Life Utilization of Insurance-Linked Securities (2010).
[33] Partner Re, *A Balanced Discussion on Insurance Linked-Securities* (2008), available at http://www.parterre.com.
[34] Christopher Kampa, Alternative Risk Transfer: The Convergence of the Insurance and Capital Markets, Part II, Non-Life Utilization of Insurance-Linked Securities (2010).

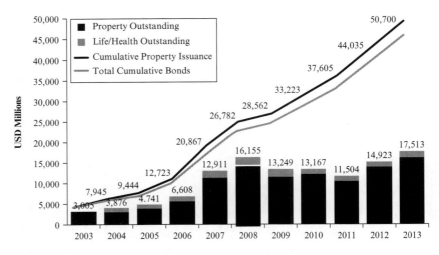

Source: Aon Benfield Securities, Insurance-Linked Securities: Capital Revolution – ILS Market Expands to New Heights (2013), available at http://thoughtleadership.aonbenfield. com/Documents/20130830_ab_ils_annual_report_2013.pdf.

Figure 6.3   Outstanding and cumulative catastrophe bond volume, 2003–2013 (years ending June 30)

occupancy, coverage, and business.[35] A parametric trigger is based on parameters associated with a peril, such as event location and intensity.[36] For an earthquake, the parametric trigger might be the magnitude of the earthquake. For a hurricane, the parametric trigger might be the location of landfall or the average sustained wind speed.

The first catastrophe bond issued by USAA in 1997 was triggered by indemnity, as it is based on ultimate net loss defined under USAA's portfolio parameters (e.g., cover under existing policies/renewals and new policies in 21 listed states). A Cayman SPR vehicle, Residential Re, was established to write the reinsurance contract to USAA and issued bonds to investors. The bond was structured to give the insurer coverage to a maximum of $500 million at 80 percent participation (e.g., 20 percent coinsurance), equivalent to $400 million of reinsurance cover. This hur-

---

[35]   Christopher Kampa, Alternative Risk Transfer: The Convergence of the Insurance and Capital Markets, Part II, Non-Life Utilization of Insurance-Linked Securities (2010).

[36]   Christopher Kampa, Alternative Risk Transfer: The Convergence of the Insurance and Capital Markets, Part II, Non-Life Utilization of Insurance-Linked Securities (2010).

Table 6.1   Cross-comparison of catastrophe bond triggers

	Indemnity	Index	Parametric
Trigger	Based on insurer's actual losses	Based on industry loss estimates	Based on defined parameters of catastrophic event
Advantages	Eliminate basis risk	More transparent; eliminates moral hazard risk	More transparent; quicker settlement period leading to increased liquidity
Disadvantages	Long recovery period to calculate loss claims, leading to less liquidity for investors; potential for moral hazard risk	Exposes risk transferor to basis risk	Exposes risk transferor to basis risk

Note: LIBOR is the abbreviation for "London interbank offered rate". See David Cummins, *CAT Bonds and Other Risk-Linked Securities: State of the Market and Recent Developments*, 11 Risk Management and Insurance Review 23 (2008).

ricane bond set the stage for many others to follow.[37] While most of the earliest bonds in the market featured indemnity triggers, a gradual shift towards parametric and index triggers has occurred since 2000, partly as a result of the lack of transparency surrounding portfolios of risk, which investors find difficult to analyse. Table 6.1 shows the strengths and weaknesses of the three triggers, which can partly explain the shift.[38]

C.   Reasons for Insurance Securitization

Several factors have combined to accelerate the development of insurance securitization. The first is the increase of value-at-risk, such as the increased concentration of the world's population and property in vulnerable areas, combined with the fact that the capacity of the insurance industry to cover catastrophic losses is insufficient. As mentioned in the introduction, data show that the frequency of and losses from natural

---

[37]   Erik Banks, Catastrophic Risk, Analysis and Management 113 (2005).
[38]   In Table 6.1, "basis risk" is the risk that the index calculation will not be the same as the client's actual loss. How moral hazard could occur will be discussed in the next section.

catastrophes have mounted, and there was also a marked increase in the amount of insured losses from 1970 to 2014.[39] For example, Hurricane Katrina, which caused insured losses of $119 billion in 2005, provided an impetus for insurers to develop ARTs to cover catastrophe risk, since such losses are very large relative to the total equity capital of global reinsurers but represent less than 1 percent of the value of US capital markets.[40] Insurance-linked securities could provide additional sources of capital and enable insurers to provide coverage for catastrophe losses even when these high losses are expected.[41]

The second factor is the development of catastrophe models. With the steadily growing body of data on catastrophe events, catastrophe model firms have developed sophisticated predictive models, and natural catastrophe risk is evolving away from a highly uncertain line of business.[42] Technologies enhance market transparency and facilitate the creation of complex insurance-linked securities.[43]

The third factor is the contrast with reinsurance. Traditionally, insurers have protected themselves through private reinsurance contracts whereby portions of their losses from a catastrophic disaster are covered by some type of treaty or excess-loss arrangement.[44] Reinsurance provides insurers with risk financing, while insurance-linked securities provide insurers with risk transfer.[45] Insurance-linked securities are regarded as useful tools to transfer catastrophe risk to the capital markets and serve as a complement to traditional reinsurance.

---

[39]   Swiss Re Sigma, *Natural Catastrophes and Man-made Disasters in 2014: Convective and Winter Sotmrs Generate Most Losses* (2015), available at http://www.actuarialpost.co.uk/downloads/cat_1/sigma2_2015_en.pdf.

[40]   Ernst N. Csiszar, *An Update on the Use of Modern Financial Instruments in the Insurance Sector*, 32 The Geneva Papers on Risk and Insurance—Issues and Practice 319 (2007); Swiss Re, Insurance-Linked Securities Market Update, Vol. XVI (July), Zurich Switzerland (2011).

[41]   Dwight M. Jaffee and Thomas Russell, *Catastrophe Insurance, Capital Markets, and Uninsurable Risk*, 64 Journal of Risk and Insurance 205 (1997).

[42]   Swiss Re, Innovating to Insure the Uninsurable 14 (2005).

[43]   David Cummins and Pauline Barrieu, *Innovations in Insurance Markets: Hybrid and Securitized Risk-Transfer Solutions* in Handbook of Insurance 547 (G. Dionne ed., 2013).

[44]   Kenneth Froot, The Financing of Catastrophe Risk 151–152 (1999).

[45]   Christopher Kampa, Alternative Risk Transfer: The Convergence of the Insurance and Capital Markets, Part I, A Broad Overview (2010).

III. SUPPLY AND DEMAND FOR INSURANCE SECURITIZATION

Whether insurance-linked securities provide a sustainable solution for distributing catastrophe risk depends in large part on whether capital market investors generate sufficient demand for them. Insurance-linked securities are attractive and very easy for insurers to create, as discussed above. In fact, there is very little supply constraint here. What matters is whether capital market investors are willing to take up the securities at a price that insurers are willing to accept.

A. Supply of Insurance-linked Securities

Insurers are significant participants in the insurance-linked securities market. Because most insurance-linked securities issuances are controlled or sponsored by insurers, insurers essentially design insurance-linked securities. Insurers manage their own risk exposures through ART-related mechanisms and supply insurance-linked securities products to increase their capacity to underwrite catastrophes.[46] (Investment banks remain the primary arrangers of insurance-linked securities, given their experience in other types of securitizations and their ability to place bonds through large distribution networks.[47])

For issuers of insurance-linked securities, there are many benefits, especially in contrast with those of reinsurance. The first benefit of insurance-linked securities is that the securities may lower the cost to insurers at times when the reinsurance market is "hard". Although reinsurance is the mechanism by which primary insurers "unload" or "redistribute" exposure to risks that they do not wish to keep in their portfolios, reinsurance underwriting confronts a cycle that goes through alternating phases of "hard" and "soft" markets. As discussed in Chapter 5, the existence of hard markets leads to decreased supply but increased premiums.[48] However, reinsurance-underwriting cycles have a low correlation with

---

[46] E. Banks, Alternative Risk Transfer: Integrated Risk Management through Insurance, Reinsurance, and the Capital Markets 53 (2004).

[47] E. Banks, Alternative Risk Transfer: Integrated Risk Management through Insurance, Reinsurance, and the Capital Markets 118 (2004).

[48] A "hard market" typically refers to the time period soon after the occurrence of a catastrophe event, when reinsurers generally limit their coverage and charge higher premiums for these risks. See David Cummins and Olivier Mahul, Catastrophe Risk Financing in Developing Countries: Principles for Public Intervention 194 (2009). See further discussion in Chapter 5.

securities market returns.[49] During hard-market periods, insurance-linked securities can reduce the costs associated with most severe types of catastrophic risks.[50] In that regard, the presence of insurance-linked securities as an alternative way of transferring catastrophe risk can help prevent reinsurance prices from increasing faster than expected and avoid the effects of such underwriting cycles.[51]

A second benefit of insurance-linked securities is that they have the potential to solve capacity problems in insurance markets. Because of the huge size of capital markets and the potential for pure risk transfer, insurance-linked securities help solve capacity gaps.[52] Representatives from one insurance company stated that their company was not able to purchase the amount of reinsurance needed in this risk category from traditional reinsurers. As a result, the company paid premiums to SPVs to get the coverage via catastrophe bonds, as well as to replace some of the company's reinsurance coverage in this risk category.[53] The potential capacity of capital markets creates greater coverage capacity than traditional risk pooling can provide.[54]

A third benefit of insurance-linked securities is that these bonds provide

---

[49]   David Cummins and Pauline Barrieu, *Innovations in Insurance Markets: Hybrid and Securitized Risk-Transfer Solutions* in Handbook of Insurance 547 (G. Dionne ed., 2013).

[50]   US Government Accountability Office, Catastrophe Insurance Risk: The Role of Risk-Linked Securities and Factors Affecting Their Use, GAO-02-941 (2002).

[51]   US Government Accountability Office, Catastrophe Insurance Risks: Status of Efforts to Securitize Natural Catastrophe and Terrorism Risk, GAO-03-1033 (2003).

[52]   Dwight Jaffee and Thomas Russell, *Catastrophe Insurance, Capital Markets, and Uninsurable Risks*, 64 Journal of Risk and Insurance 205 (1997); Sylvie Bouriaux and William L. Scott, *Capital Market Solutions to Terrorism Risk Coverage: A Feasibility Study*, 5 Journal of Risk Finance 34 (2004) (noting that risk transfer to the capital markets can increase coverage capacity by taking pressure off insurers to maintain large capital surpluses). See also Andrew Gerrish, *Terror CATs: TRIA's Failure to Encourage a Private Market for Terrorism Insurance and How Federal Securitization of Terrorism Risk May Be a Viable Alternative*, 68 Washington and Lee Law Review 1825 (2011).

[53]   US Government Accountability Office, Catastrophe Insurance Risks: Status of Efforts to Securitize Natural Catastrophe and Terrorism Risk, GAO-03-1033 (2003).

[54]   Jaffee and Russell point out that the losses from a major hurricane might be $100 billion. However, the stock market drops in value by more than that, and nobody bats an eyelid. See Dwight Jaffee and Thomas Russell, *Catastrophe Insurance, Capital Markets, and Uninsurable Risks*, 64 Journal of Risk and Insurance 205 (1997). See also Sylvie Bouriaux and William L. Scott, *Capital*

multiyear protection, while reinsurance is often limited to a 12-month time frame.[55] Multiyear protection insulates insurers against cyclical price changes inherent in catastrophe risk and reduces transaction costs.[56] As a result of these benefits, insurers prefer catastrophe bonds with three- or five-year maturity dates, so insurers can respond to new information, protect against reinsurance price changes, and keep costs down.[57]

A fourth benefit of insurance-linked securities is that they reduce the exposure to the credit risk of an individual reinsurer. As with any financial transaction, there is a potential credit risk associated with reinsurance due to the risk of insolvency or slow payment by the reinsurer.[58] Catastrophe bonds provide full collateralization of losses by entering into a swap agreement with the swap counterparty, which eliminates credit risk for clients.[59]

Insurance-linked securities, however, are not without their constraints vis-à-vis traditional reinsurance for issuers, and these constraints may prevent insurance-linked securities from becoming more prominent. The first constraint of issuance of insurance-linked securities is that the total costs are high in comparison to reinsurance, at least when reinsurance is experiencing a soft market offering coverage to insurers on reasonable terms. These costs include legal fees, broker fees, rating agency fees, bank fees, actuarial/modeling fees, administrative costs, and relatively high rates of return paid to investors. In contrast, reinsurance has only brokerage fees.[60] Some insurance company officials have estimated that the total costs associated with catastrophe bonds could be double that of traditional reinsurance.[61] The cost of insurance-linked securities is mostly

---

Market Solutions to Terrorism Risk Coverage: A Feasibility Study, 5 Journal of Risk Finance 34 (2004).

[55] Christopher Kampa, Alternative Risk Transfer: The Convergence of the Insurance and Capital Markets, Part I, A Broad Overview (2010).

[56] David Cummins, *CAT Bonds and Other Risk-Linked Securities: State of the Market and Recent Developments*, 11 Risk Management and Insurance Review 23 (2008).

[57] David Cummins, *CAT Bonds and Other Risk-Linked Securities: State of the Market and Recent Developments*, 11 Risk Management and Insurance Review 23 (2008).

[58] Partner Re, *A Balanced Discussion on Insurance Linked-Securities* (2008), available at http://www.parterre.com.

[59] See Figure 6.1 above. There can still be counterparty risk, however, as the Lehman Brothers bankruptcy demonstrated.

[60] Partner Re, *A Balanced Discussion on Insurance Linked-Securities* (2008), available at http://www.parterre.com.

[61] US Government Accountability Office, Catastrophe Insurance Risks: Status of Efforts to Securitize Natural Catastrophe and Terrorism Risk, GAO-03-1033 (2003).

driven by the high premiums required to attract investors, since the yields on catastrophe bonds have generally exceeded the yields on some high-yield corporate debt.[62] The cost of insurance-linked securities is also partly driven by administrative and transaction costs, since these can be highly customized instruments (for example, the development of parametric triggers), particularly the first time they are issued.[63] However, these costs are likely to fall over time as the bonds gain acceptance with the public, and as competition in the market for issuer-assistance improves. Some of the high initial expenses of issuing bonds could be decreased or at least spread over multiple issuances through a shelf offering.[64]

The second constraint of issuance of insurance-linked securities is that they expose the insurer to basis risk, while reinsurance does not. Basis risk is the risk that the index calculation will not be the same as the insurer's actual loss.[65] Basis risk generally reflects the possibility that insurance-linked securities may not be partially or fully triggered (for covered perils) even when the sponsor of the insurance-linked securities has suffered a loss.[66] Since most insurance-linked securities' structures are index-based, basis risk can be significant and may result in the bond not paying enough to cover the insurer's losses.[67]

B.   Demand for Insurance-linked Securities

Investors provide the capital, or risk capacity, that produces the demand for insurance-linked securities. These investors are vital to the ART market, as they allow risks to be assumed, transferred, hedged, or otherwise transformed. In fact, without this capital, the insurance-linked securities market would simply cease to function. Investors are generally

---

[62]   US Government Accountability Office, Catastrophe Insurance Risks: Status of Efforts to Securitize Natural Catastrophe and Terrorism Risk, GAO-03-1033 (2003).

[63]   US Government Accountability Office, Catastrophe Insurance Risks: Status of Efforts to Securitize Natural Catastrophe and Terrorism Risk, GAO-03-1033 (2003).

[64]   Partner Re, *A Balanced Discussion on Insurance Linked-Securities* (2008), available at http://www.parterre.com.

[65]   Partner Re, *A Balanced Discussion on Insurance Linked-Securities* (2008), available at http://www.parterre.com.

[66]   Best's Methodology and Criteria, *Gauging the Basis Risk of Catastrophe Bonds* (2017), available at http://www3.ambest.com/ambv/ratingmethodology/OpenPDF.aspx?rc=197690.

[67]   Partner Re, *A Balanced Discussion on Insurance Linked-Securities* (2008), available at http://www.parterre.com.

large institutions that seek adequate returns on their investment portfolios in return for the provision of capital.[68] Many constraints impede the demand for investors to purchase the insurance-linked securities that insurers are willing to offer them. Because most trading to date has taken place through catastrophe bonds, those instruments will be the focus of the following discussion.

The first constraint arises from the fact that the risks of catastrophe bonds are difficult to assess. Data about natural perils and geographical areas are very important for the actuarial calculations of loss; however, there are still no standardized and united data.[69] Perhaps partly for that reason, a mutual fund industry official stated that mutual fund companies had not purchased catastrophe bonds in funds potentially available to individual investors because the companies were not capable of evaluating the risks.[70]

The second constraint lies in the costly and time-consuming nature of due diligence by investors given catastrophe bonds' complicated structures. It is not cost-effective for investors to improve their technical capability to analyse the risks of catastrophe bonds, because those bonds are different from the securities in which they currently invest.[71]

The third constraint lies in the limited liquidity of insurance-linked securities. Some large mutual fund representatives share the concern that catastrophe bonds are relatively illiquid compared to traditional bonds and equities.[72] Even investors who have already purchased catastrophe bonds say that they have limited their investments in the bonds to no more than 3 percent of their total portfolios.[73] Given the few investors who are interested in catastrophe bonds, 3 percent is not a high number in their investment strategies.

---

[68] E. Banks, Alternative Risk Transfer: Integrated Risk Management through Insurance, Reinsurance, and the Capital Markets 57 (2004).

[69] Christopher Kampa, Alternative Risk Transfer: The Convergence of the Insurance and Capital Markets, Part II, Non-Life Utilization of Insurance-Linked Securities (2010).

[70] US Government Accountability Office, Catastrophe Insurance Risk: The Role of Risk-Linked Securities and Factors Affecting Their Use, GAO-02-941 (2002).

[71] US Government Accountability Office, Catastrophe Insurance Risks: Status of Efforts to Securitize Natural Catastrophe and Terrorism Risk, GAO-03-1033 (2003).

[72] US Government Accountability Office, Catastrophe Insurance Risks: Status of Efforts to Securitize Natural Catastrophe and Terrorism Risk, GAO-03-1033 (2003).

[73] US Government Accountability Office, Catastrophe Insurance Risks: Status of Efforts to Securitize Natural Catastrophe and Terrorism Risk, GAO-03-1033 (2003).

The fourth constraint lies in the adverse impact of the 2008 financial crisis on the creditworthiness of these securities. As Figure 6.1 above illustrates, to ensure that the necessary interest rates are obtained, the trustee typically enters into a swap agreement in which the swap counterparty provides protection to the trust if the interest obtained falls below the amount necessary to pay the required interest payments to the various tranche investors.[74] Prior to the credit crisis, catastrophe bonds were structured with a total return swap counterparty, which was usually an investment bank, to guarantee the collateral pool backing the bonds.[75] Investment banks often acted as the swap counterparty in these transactions.[76] In the wake of the financial crisis, investors lost faith in the creditworthiness of the underlying guarantees by investment banks, which impaired the market for catastrophe bonds.[77]

Nevertheless, insurance-linked securities are still attractive for investors since they have little correlation with the risks that traditional securities face. Most natural catastrophes are uncorrelated with economic conditions, and investments in catastrophic risk could therefore help market investors diversify their portfolios.[78] Insurance-linked securities are a desirable addition to an investor's portfolio to achieve portfolio diversification. For example, some pension funds purchase catastrophe

---

[74]   See E. Banks, Alternative Risk Transfer: Integrated Risk Management through Insurance, Reinsurance, and the Capital Markets 125–126 (2004).

[75]   James S. Gkonos and Braden A. Borger, *At the Crossroads of Insurance and the Capital Markets: An Analysis of the Current Regulatory Environment and New Developments in Alternative Risk Transfers*, New Appleman on Insurance: Current Critical Issues in Insurance Law (2010).

[76]   Lehman Brothers often acted as the swap counterparty in these transactions. In 2008, the collapse of Lehman Brothers triggered substantial credit-related losses due to the underlying swap arrangements that had been designed to protect investors against counterparty risk. See James S. Gkonos and Braden A. Borger, *At the Crossroads of Insurance and the Capital Markets: An Analysis of the Current Regulatory Environment and New Developments in Alternative Risk Transfers*, New Appleman on Insurance: Current Critical Issues in Insurance Law (2010).

[77]   Christopher Kampa, Alternative Risk Transfer: The Convergence of the Insurance and Capital Markets, Part II, Non-Life Utilization of Insurance-Linked Securities (2010).

[78]   Kenneth Froot, The Financing of Catastrophe Risk 257 (1999). Generally speaking, investors, especially institutional investors, attempt to invest in equities and debt from a wide range of companies, industries, and geographic locations to minimize their exposure to any particular risk in the event of an economic downturn. See US Government Accountability Office, Catastrophe Insurance Risks: Status of Efforts to Securitize Natural Catastrophe and Terrorism Risk, GAO-03-1033 (2003).

bonds because they are uncorrelated with other credit risks in their bond portfolios and help diversify their investment risks.[79]

C.   A Short Conclusion

There are challenges for the insurance-linked securities market. The constraints associated with catastrophe bonds, especially the demand-side of insurance-linked securities, are real. These may be the reasons why insurance-linked securities have not developed as quickly as predicted and are still in their infancy. The insurance-linked securities market is relatively new, and the rate of growth is impressive (see Figures 6.2 and 6.3 above), albeit from a very small base. More insurers—the suppliers—are participating in the catastrophe bond market.[80] The market has seen participation not only from traditional insurance and reinsurance companies but also from other noninsurance companies, such as Tokyo Disney and Universal Studios.[81] From the perspective of investors, catastrophe bonds have yielded above-average returns in the past while facilitating the diversification of portfolios.[82] Investor demand for catastrophe bonds is now expected to grow. This suggests that further growth of the catastrophe bond market is possible. After the financial crisis, the insurance-linked securities market is coming back, and its future is promising.

IV.   IS SECURITIZATION OF CATASTROPHE RISK FEASIBLE IN CHINA?

A.   Feasibility of Securitization of Catastrophe Risk

The feasibility of securitization of risk in China depends significantly on the supply-demand dynamics in which the insurance industry creates the

---

[79]   US Government Accountability Office, Catastrophe Insurance Risks: Status of Efforts to Securitize Natural Catastrophe and Terrorism Risk, GAO-03-1033 (2003).

[80]   Aon Benfield Securities, *Insurance-Linked Securities—Consistency and Confidence* (2011), available at http://thoughtleadership.aonbenfield.com/Documents/201108_ab_securities_ils_annual_2011.pdf.

[81]   Christopher Kampa, Alternative Risk Transfer: The Convergence of the Insurance and Capital Markets, Part II, Non-Life Utilization of Insurance-Linked Securities (2010).

[82]   Christopher Kampa, Alternative Risk Transfer: The Convergence of the Insurance and Capital Markets, Part II, Non-Life Utilization of Insurance-Linked Securities (2010).

supply for insurance-linked securities and the capital market generates the demand for those securities.

1. Supply

The supply by the insurance industry for insurance-linked securities in China depends on numerous conditions. Some conditions are favorable, while others are not.

(a)   Positive conditions First, China is developing a system of catastrophe insurance, and that system will likely give rise to a greater demand for insurance-linked securities on the part of insurers. Local catastrophe insurance pilot programs were launched in Shenzhen City, Ningbo City, and Chuxiong Region in 2014. Due to the nature of catastrophe insurance discussed above and the advantages of securitization of risk products compared to reinsurance, insurance companies will have an incentive to issue such securities or bonds in the capital market.

A second positive condition lies in the rapid development of the Chinese insurance market and insurance companies, some of which are big enough to explore and issue insurance-linked securities. From 2001 to 2012, China was one of the fastest growing insurance markets in the world and in 2012 stood at $249 billion, representing 5.5 percent of the world's total insurance premiums (up from 1 percent in 2001).[83]

A third positive condition lies in the international cooperation with foreign insurance companies. Securitization of catastrophe insurance risk has developed in the European Union and in the United States over a period of decades. Their experiences with securitization institutions and technologies can inform Chinese insurance companies, especially those that work with foreign reinsurers. For example, Swiss Re has rich experience with insurance-linked securities, and it was one of the first foreign reinsurers to enter China after the opening of China's insurance markets. Decades ago, Swiss Re set up representative offices in Beijing and Shanghai, and opened its Beijing branch to conduct business at the end of 2003.

A fourth positive condition lies in growing catastrophe data, which helps the pricing of insurance-linked securities. The price of insurance-linked securities is a major determinant of supply,[84] and the reasonableness of insurance-linked securities' prices determines the success of insurance-

---

[83]   Aon Benfield, *China P&C Insurance and Reinsurance Market Report* (2013), available at http://thoughtleadership.aonbenfield.com/documents/20130923_ana lytics_china_market.pdf.
[84]   Sylvie Bouriaux and Richard MacMinn, *Securitization of Catastrophe Risk: New Developments in Insurance-linked Securities and Derivatives* (March

linked securities. The pricing of insurance-linked securities is partly based on the available data about frequency and scope of catastrophe events. From 1998 to 2002, Swiss Re, cooperating with Beijing Normal University, completed the Digital Map of China Catastrophe Events, which includes historical data on geography, weather, and so on, from the twelfth century onwards.[85] This digital map has been very helpful for the pricing of insurance-linked securities. In 2009, the National Disaster Reduction Centre of China, along with Swiss Re and Munich Re, developed a new catastrophe zoning system and improved data transparency as a vital step to sustainable management of natural catastrophe risk in China.[86] In the same year, the China Insurance Regulation Commission promulgated a Catastrophe Insurance Data Acquisition Regulation (JR/T0054-2009), which encourages insurance companies to collect catastrophe data and sets up standards for acquisition of such data.[87] The provisions of this Regulation substantially improved the precision of data entry, and the efficiency and accuracy of data exchange.

(b)　Negative conditions A first negative condition lies in the high upfront costs of insurance-linked securities, which include legal fees, broker fees, rating agency fees, bank fees, and actuarial/modeling fees.[88] Insurance-linked securities are still very new in China, and insurance companies, intermediaries, and investors are not familiar with them. Therefore, upfront costs may be higher than expected. If the price of the insurance-linked securities is higher than traditional reinsurance, its competitiveness will be weak.

A second negative condition lies in the illiquidity of insurance-linked securities. Catastrophe bonds, for example, are relatively illiquid compared with traditional bonds and equities.[89] The limited liquidity of

---

3, 2012), available at http://www.macminn.org/papers/Securitization%20of%20 Catastrophe%20Risk.pdf.

[85]　Xi Guo and Xinjiang Wei, *The Difficulties and Solutions for Issuing Catastrophe Bonds in China*, 6 China Insurance 22 (2005).

[86]　Swiss Re, *Swiss Re Promotes CRESTA's New Zoning for Better Natural Catastrophe Risk Management in China* (2009), available at http://www.swissre. com/locations/swiss_re_promotes_crestas_new_zoning_for_better_natural_catast rophe_risk_management_in_china.html.

[87]　CIRC Order 52 (2009).

[88]　Christopher Kampa, Alternative Risk Transfer: The Convergence of the Insurance and Capital Markets (2010).

[89]　US Government Accountability Office, Catastrophe Insurance Risks: Status of Efforts to Securitize Natural Catastrophe and Terrorism Risk, GAO-03-103 (2003).

insurance-linked securities makes them less attractive to investors in the secondary market.

A third negative condition lies in the immaturity of service agencies, e.g., investment banks. Insurance-linked securities require trustworthy "outside" information in order to succeed. Someone has to compile statistics on the occurrence and magnitude of catastrophes so that contracts can then be linked to these quantities. (Clearly, insurers themselves cannot perform this role, since they would have an incentive to distort the data to maximize their own returns.[90]) Some institution—presumably a respected governmental agency with a reputation for integrity and neutrality—needs to be responsible for collecting data, establishing a catastrophe risk model, and appraising the possibility of certain catastrophes in specific areas and predicting losses. However, such institutions or agencies are not established or conducting business in China.

Some of these negative conditions are present both in China and in other countries. These obstacles to the more widespread use of insurance-linked securities are not insurmountable, since many countries have already issued insurance-linked securities. As a result of the lack of a sophisticated legal framework in China, insurance companies may have the alternative choice to issue insurance-linked securities abroad. From this perspective, the positive conditions are promising. With the development of catastrophe insurance pilot programs, insurers will have more incentives to cede catastrophe risk through alternative risk mechanisms, and the advent of insurance-linked securities could be expected. Indeed, Panda Re, which is registered as a Bermuda-domiciled SPV and sponsored by China Re, issued catastrophe bonds covering earthquakes in China to the international capital market in 2015.[91]

2. Demand

Capital markets generate the demand for insurance-linked securities, so it is necessary to survey the current conditions and the future of capital markets in China. Unfortunately, capital markets in China do not have sufficient incentives to meet the demand of insurance-linked securities, at least at this point in time.

Investors have the final say over the demand for insurance-linked securities. In the United States, the European Union, and other advanced

---

[90]   Miao Miao, *Empirical Research on Catastrophe Risk Securitization*, 205 Group Economy 188 (2006).

[91]   ARTEMIS, Panda Re Cat Bond an "Important Breakthrough" for China Re (2015), available at http://www.artemis.bm/blog/2015/07/02/panda-re-cat-bond-an-important-breakthrough-for-china-re/.

capital markets, institutions are the main investors, especially for sophisticated products such as insurance-linked securities. In contrast, in China, individual investors hold a large portion of all assets, and institutional investors are relatively less important. Compared to institutional investors, individual investors have less skill and less incentive to purchase insurance-linked securities as portfolio diversification due to the complicated structure and complex transaction process of these securities. An alternative choice for Chinese insurers is to tap institutional investors to do issuance abroad, where institutional investors actively purchase offerings.

In addition, the perverse effects of securitization exposed during the 2008 financial crisis decreased incentives for investors to purchase insurance-linked securities, since asset-backed securitization served as a good model for catastrophe risk securitization due to their similarity in nature in practical experiences. In the wake of the financial crisis, investors lost faith in the creditworthiness of the underlying guarantees by investment banks, which impaired the insurance-linked securities market.[92] Asset-backed securitization is still in its initial stages in China, and even more so for insurance-linked securities. The fear of investing in asset-backed securities might "infect" the investment for insurance-linked securities. Under such circumstances, it is not easy for insurance-linked securities to attract investors.

Last but not least, the lack of a mature legal and regulatory framework for insurance-linked securities not only affects supply but also demand, since it creates uncertainty as to investment returns and the ability to recover for securities fraud. In China's existing regulatory framework, the approval and regulation of insurance-linked securities are still not clear, and relevant laws are nonexistent. Like most capital market transactions, insurance-linked securities require the backing of a sophisticated legal framework. The absence of such a framework will substantially hinder the issuance and purchase of insurance-linked securities.

B. Catastrophe Bonds in China

Catastrophe bonds are the most prominent form of insurance-linked securities, and if insurance-linked securities are introduced in the near future, it is likely that they will take the form of catastrophe bonds. Catastrophe bonds are relatively well developed in the United States, the European

---

[92] Kurt Eggert, *The Great Collapse: How Securitization Caused the Subprime Meltdown*, 41 Connecticut Law Review 1257 (2008).

Union, and other countries, and so many good practices can be emulated. Buyers of catastrophe bonds are typically institutional investors, rather than individuals, and are sophisticated enough to undertake the risk of such a new financial product. Moreover, the issuance of catastrophe bonds is not so complicated—compared to other insurance-linked securities—that insurance companies as the issuers cannot grasp the technology in a short period of time. In this section, some of the legal obstacles that confront the issuance of catastrophe bonds in China are described.

An SPV or a Special Purpose Entity serves as the issuer that transfers catastrophe risks from insurers to investors. An SPV is usually set up by the insurance company or reinsurance company to issue the catastrophe bonds.[93] However, choosing the form of an SPV poses problems under China's legal system. In the United States, an SPV can take the form of a limited liability company newly created for a specific purpose, a trust, or an existing corporation.[94] Such options are not available in China, however. If an SPV is set up as a corporation, according to the current Chinese Company Law, the issuance of debentures or bonds requires net assets of a limited liability company of at least RMB 60 million, while the cumulative value of debentures or bonds must not exceed 40 percent of the company's net assets.[95] Such a high threshold prevents SPVs from being set up as corporations. In the United States, SPVs are generally set up as trusts, not as corporations. More importantly, the legal nature of an SPV—whether it is an insurance agency or a securities agency—is not clear under Chinese law. Assuming the SPV is securities agency, there is no relevant regulatory provision authorizing the China Securities Regulation Commission to regulate its business and products.[96] Assuming the SPV is an insurance agency, it may violate the provisions of the Insurance Company Equity Management Measures, Article 5 of which provides that if an insurance company controls another insurance company, these two companies cannot conduct similar business.[97] Since the SPV and its sponsor—the ceding company—both conduct catastrophe insurance busi-

---

[93]   Partner Re, *A Balanced Discussion on Insurance Linked Securities* (2008), available at http://www.parterre.com.

[94]   Jian Wang, *What China Can Learn from US Securitization*, 25 International Finance Law Review 22 (2006).

[95]   Company Law (PRC), which is promulgated by Standing Comm. National People's Congress, October 27, 2005, effective January 1, 2006, Article 161 (2006), available at http://www.gov.cn/ziliao/flfg/2005-10/28/content_85478.htm.

[96]   Tao Liang, *Analysis of Legal Hurdles to Issuing Catastrophe Bonds in China* (2014), available at http://works.bepress.com/tim_liang/34/.

[97]   Tao Liang, *Analysis of Legal Hurdles to Issuing Catastrophe Bonds in China* (2014), available at http://works.bepress.com/tim_liang/34/.

ness and the transfer of catastrophe risks, the SPV's operation could not be allowed by the regulator.

Issuing catastrophe bonds through offshore SPVs[98] may be a practical choice for avoiding some of the difficulties with the legal framework in China. Furthermore, offshore securitization has fewer obstacles and demands less of the collateral servicing system.[99] Offshore issuances will also increase the investor base. This would not only achieve the goal of transferring catastrophe risk but also provide experience in the issuance of catastrophe bonds and other insurance-linked securities in the Chinese domestic market. Panda Re has been a very successful first step in using an offshore SPV to transfer earthquake risk in China.[100]

V.   CONCLUSION

Securitization of catastrophe risk is a creative invention based on the highly developed insurance industry and capital market. Due to institutional shortcomings, including legal framework, regulatory institutions, and so on, issuing insurance-linked securities will still take a long time and will be a difficult process. Although there have been many criticisms of insurance-linked securities, especially during and after the financial crisis, they have still grown in recent years and are likely to continue to grow in the coming years as the impacts of climate change become evident.[101] What we should learn from the 2008 financial crisis is how to develop and regulate insurance-linked securities markets better than before. Insurance-linked securities are attracting more and more attention in the world as a result of their advantages in solving problems of catastrophe insurance and covering losses caused by disasters.[102]

---

[98]   Such as in Bermuda, Cayman Islands, Hong Kong, etc.

[99]   Wensheng Yang and Jiafeng An, *A Study on the Feasibility of China's Insurance Risk Securitization—The Obstacles for China's Insurance Risk Securitization and Related Institutional Arrangement Viewing from Catastrophe Bond*,135 Journal of Henan Institute of Financial Management 25 (2007).

[100]   Matthew Lerner, *Panda Re Catastrophe Bond Covers China Risks* (2015), available at http://www.businessinsurance.com/article/20150702/NEWS06/150709961.

[101]   Erwann Michel-Kerjan et al., *Catastrophe Financing for Governments: Learning from the 2009–2012 MultiCat Program in Mexico*, OECD Working Papers on Finance, Insurance and Private Pensions, No. 9, (OECD, 2011).

[102]   Even in middle-income countries, for example, Mexico cooperated with the World Bank and issued the CAT bond *MultiCat Mexico 2009* in October 2009. Moreover, while Chile decided in 2008 not to pursue this path, the combination of a major 8.8 earthquake in the Maulé region in February 2010 and the election of

On the other hand, China's securities market and legal framework are not yet ready for the issuance of insurance-linked securities. In fact, even in the European Union, catastrophe bonds have not developed as quickly as was predicted, and these products are still in their infancy in Europe.[103] However, policymakers in China should not overlook the bright future of securitization of catastrophe risk. It is only a matter of time, and we should be well prepared for the coming of insurance-linked securities.

---

a new administration have triggered renewed attention to developing insurance-linked securities. See Erwann Michel-Kerjan et al., *Catastrophe Financing for Governments: Learning from the 2009–2012 MultiCat Program in Mexico*, OECD Working Papers on Finance, Insurance and Private Pensions, No. 9 (OECD, 2011).

[103]   Véronique Bruggeman, Compensating Catastrophe Victims: A Comparative Law and Economics Approach 552–553 (2010).

7. Roadmap for transitional reform in China

I. INTRODUCTION

Currently, the Chinese mechanism for managing catastrophe risks or challenges is the "Whole-Nation System" (*Juguo tizhi*), which entails neither private insurance nor social insurance, but rather a kind of emergency-driven disaster relief system.[1] This book has explored and examined the important but neglected question of how to enlist private- and public-sector insurance to mitigate climate change risks and compensate climate disaster losses in China. Promoting insurance to combat climate change risks faces a number of general challenges no matter what the country. Underwriting catastrophe insurance faces both supply-side and demand-side barriers. On top of these general challenges, China faces additional hurdles, which are specific to the Chinese context, in deploying insurance successfully to fight climate change.

While there are no perfect answers, this chapter seeks to provide a roadmap to navigating these challenges and a proposed framework for tapping into private insurance as well as the public sector to address global warming and weather-related disasters in China.

II. SOLUTIONS FOR SOLVING THE LACK OF SUPPLY FOR CATASTROPHE INSURANCE

A. Behavioral Characteristics of Insurers

There is growing evidence that insurance companies often deviate from the ideal benchmark supply model for several reasons stemming from behavioral factors. Climate risk is an ambiguous and uncertain risk, and the ambiguities associated both with the probability of an extreme event occurring and the resulting outcomes raise a number of challenges for

---

[1] See Chapter 1.

insurers with respect to pricing their policies and underwriting behaviors. In evaluating catastrophe risks, insurance underwriters exhibit biases or use simplified decision rules,[2] and their behaviors often follow a safety-first type model rather than the benchmark model of maximizing expected profit.[3]

Due to the highly correlated and aggregated losses from catastrophe exposures, insurers lack the capacity and willingness to cover such losses. The supply of catastrophe insurance is limited and volatile. Outside capital is needed to help fill insurers' "capacity gap" and supplement their coverage.

B.   Expanding the Supply of Catastrophe Insurance

There are several techniques that could be used to expand the supply of catastrophe insurance in China. These include multiyear insurance, mandatory insurance, (international) reinsurance, and insurance-linked securities (ILS).

Multiyear insurance Multiyear insurance refers to insurance in which policies are sold for consecutive years rather than only for one year. This system helps solve, to some extent, supply-side challenges since it broadens insurers' capacity to cover losses by extending the term of an insurance policy and the expected premiums over the life of the policy. When a catastrophe occurs, a multiyear policy allows the insurer to spread the associated losses over multiple years. Moreover, multiyear insurance can reduce transaction costs for both insurers and consumers.

Mandatory insurance Mandatory insurance can also, to some extent, help to solve the supply problem since it can help reduce the number of insureds who are adverse to taking out insurance cover. By forcing universal participation in a risk pool, mandatory insurance prevents lower-risk groups from opting out of the pool, and in turn attracts insurers to provide coverage. Of course, mandatory insurance arrangements need outside capital to help fill insurers' "capacity gap", which will be discussed below.

Reinsurance Reinsurance is another tool to help solve the supply challenge since insurers need greater financial capacity when underwriting catastrophic risks. Private reinsurance may help expand this capacity. In

---

[2]   L. Cabantous, D. Hilton, H. Kunreuther and E. Michel-Kerjan, *Is Imprecise Knowledge Better than Conflicting Expertise? Evidence from Insurers' Decisions in the United States*, 42 Journal of Risk and Uncertainty 211 (2011).

[3]   Howard Kunreuther and Erwann Michel-Kerjan, *Economics of Natural Catastrophe Risk Insurance* in Handbook of the Economics of Risk and Uncertainty 651 (Mark J. Machina and W. Kip Viscusi eds., 2014).

addition, government-sponsored reinsurance may allow the government to act as a last resort for primary insurers, expanding their capacity to write insurance and thus encouraging its supply.

Insurance risk securitization Insurance risk securitization is the last but not least important tool for enhancing the supply of catastrophe insurance coverage by increasing the capacity of private insurers. Through insurance-linked securities (ILS), risks can be stripped and transferred, creating a derivative security representing the amount and type of risk that capital market investors are prepared to underwrite (for a price). This market is underdeveloped in China but could be an important solution for expanding supply in the long run.

III.   SOLUTIONS FOR SOLVING THE LACK OF DEMAND FOR CATASTROPHE INSURANCE

A.   Behavioral Characteristics of Insureds

Homeowners' decisions not to purchase catastrophe insurance are partly due to factors such as not perceiving the hazard to be a serious problem, and partly due to homeowners' misunderstandings about the nature of insurance. Disasters are normally low-probability high-consequence (LP-HC) events. Empirical studies have revealed that individuals often make decisions predicated on intuitive thinking bias and myopic loss aversion, which leads to ignorance of a potential disaster.[4] Consumers ignore such catastrophe risk because they believe "it will not happen to me". This attitude makes consumers underestimate the risks of being exposed to climate disasters and decide not to purchase sufficient insurance coverage voluntarily.

In addition, some potential victims, such as homeowners, consider premium payments for catastrophe insurance as a loss with no potential payoff. When several years pass with no insured loss, individuals may feel they have wasted money on premiums and many of them cancel their catastrophe insurance policies.[5] This results in low demand.

Further, due to cultural aspects of insurance consumption in China, many people are reluctant to discuss "accidents" or "death" ex ante,

---

[4]   Howard Kunreuther, *All-Hazards Homeowners Insurance: Challenges and Opportunities*, 21 Risk Management and Insurance Review 141 (2018).
[5]   Erwann Michel-Kerjan, Sabine Lemoyne de Forges and Howard Kunreuther, *Policy Tenure under the U.S. National Flood Insurance Program*, 32 Risk Analysis 644 (2012).

since they are afraid that doing so may induce mishaps.[6] Even worse, because of the bad reputation of insurance companies in China, many consumers lack trust in insurers to honor policies and to pay out when insured accidents occur.[7] Thus it is challenging to increase the demand for insurance in China.

Besides these behavioral anomalies, potential victims may well be acting rationally when choosing not to purchase catastrophe insurance, thanks to the availability of government bailout and relief schemes through the Whole-Nation System. Some have opined that "solidarity kills market insurance"[8] and have criticized government intervention as "catastrophic responses to catastrophic risks".[9] The low demand for insurance partially results from the ex post generosity of such government bailouts.

B.   Enhancing the Demand for Catastrophe Insurance

Risk communication Communicating risks to the public is important for better risk awareness and improved disaster preparedness. Catastrophe risk education offered by government agencies and insurers could help transform cultural attitudes concerning insurance consumption in China, and thus increase residents' risk awareness of and demand for catastrophe insurance. "More accurate information on the specific risks for individual properties in hazard-prone areas would encourage investments in cost-effective risk reduction measures and promote the purchase and retention of disaster insurance".[10]

Mandatory insurance Mandatory insurance could also help to solve the lack of demand for insurance by changing the mindsets of people who live in catastrophe-prone areas. Many individuals do not buy catastrophe insurance because they believe that they are immune to disasters and that "it cannot happen to me". Consequently, mandatory insurance

---

[6]   Ming Zhong, Zhenzhen Sun, Gene Lai and Tong Yu, *Cultural Influence on Insurance Consumption: Insights from the Chinese Insurance Market*, 3 China Journal of Accounting Studies 24 (2015).

[7]   Ming Zhong, Zhenzhen Sun, Gene Lai and Tong Yu, *Cultural Influence on Insurance Consumption: Insights from the Chinese Insurance Market*, 3 China Journal of Accounting Studies 24 (2015).

[8]   Christian Gollier, *Some Aspects of the Economics of Catastrophe Risk Insurance* in Catastrophic Risks and Insurance 13 (2005).

[9]   Richard A. Epstein, *Catastrophic Responses to Catastrophic Risks*, 12 Journal of Risk and Uncertainty 287 (1996).

[10]   Carolyn Kousky and Howard Kunreuther, *Risk Management Roles of the Public and Private Sector*, 21 Risk Management and Insurance Review 181 (2018).

arrangements can positively improve the long-term welfare of residents by compelling them to insure.

Means-tested vouchers Of course, not everyone can afford catastrophe insurance, especially low-income residents. Traditional government relief under the Whole-Nation System needs to be reshaped in order to assist poorer property owners in paying for catastrophe insurance. Government disaster assistance could take the form of means-tested vouchers (similar to the subsidies provided to lower-income households in the United States to buy private health insurance), which cover part of the cost of insurance, and would enable poorer citizens to buy catastrophe insurance without distorting the risk mitigation benefits of risk-based pricing for more affluent households and business policyholders.[11]

All-hazard insurance policy In contrast to named-peril insurance policies, like earthquake or flood insurance, an all-hazard insurance policy covers all natural hazards. Since most catastrophes are low-probability high-consequence (LP-HC) events, insureds may cancel their policies because they consider them to be a waste of money if no insured loss occurs over several years. An all-hazards insurance policy increases the probability of policyholders making a claim during the insured periods so that insureds "feel their investment paid off and decide not to cancel their policy".[12] Currently, a number of countries have property-casualty insurance policies that cover all hazards, notably Belgium, Bermuda, France, New Zealand, Spain, and the United Kingdom.

IV. CHINA'S CHOICE: A PROPOSED FRAMEWORK FOR INCREASING RISK MITIGATION AND LOSS COMPENSATION

The year 2018 marks the tenth anniversary of the 2008 Sichuan Earthquake, also known as the Wenchuan Earthquake, which was the second largest earthquake in history in terms of economic losses.[13] The earthquake displaced at least five million people, although the number could be as high as 11 million, and caused massive damage to properties

---

[11] Howard Kunreuther, *All-Hazards Homeowners Insurance: Challenges and Opportunities*, 21 Risk Management and Insurance Review 141 (2018).
[12] Howard Kunreuther, *All-Hazards Homeowners Insurance: Challenges and Opportunities*, 21 Risk Management and Insurance Review 141, 143 (2018).
[13] The largest earthquake in terms of economic losses was the 2011 Tohoku Earthquake in Japan.

and houses.[14] However, a small proportion of the victims were covered by insurance. The impact of the Sichuan Earthquake was enormous not only because of the death toll and substantial economic losses but also because it reminded policymakers of the importance of reshaping the earthquake/catastrophe insurance system, which could save lives and reduce the potential of economic losses. The remainder of this chapter will present a proposed framework for China.

A.   Mandatory Catastrophe Insurance Arrangements

Mandatory catastrophe insurance is seriously worth considering in China. As I have discussed, mandatory insurance could help solve residents' bias against purchasing disaster insurance while increasing its expected utility.[15] In addition, mandatory coverage helps to reduce adverse selection because it may prevent lower-risk groups from opting out of the pool.[16] Furthermore, compulsory insurance can help to mitigate damage losses. As Telesetsky argues, "the most important reason for mandating catastrophe risk insurance is to compel industry actors to take action under the supervision of the profit-motivated insurance industry".[17] Under a mandatory private insurance scheme, individuals who want to lower their insurance premiums would be induced to take mitigation measures.[18]

Requiring homeowners in hazard-prone areas to purchase disaster coverage would be a feasible solution to the demand anomalies in China. Currently in China, without a legal requirement, banks are reluctant to require homeowners to have catastrophe insurance coverage. Even acquiring homeowners' insurance for more ordinary risks such as theft or fire is not a condition for mortgages, loans, or other financial services, let alone catastrophe insurance. This is the case even though the People's Bank of China (i.e., the Chinese Central Bank) issued the Residential

---

[14]   Jake Hooker, Toll Rises in China Quake, *New York Times* (May 26, 2008), available at https://www.nytimes.com/2008/05/26/world/asia/26quake.html.

[15]   See inter alia Howard Kunreuther and Mark Pauly, *Rules rather than Discretion: Lessons from Hurricane Katrina*, 33 Journal of Risk and Uncertainty 101 (2006).

[16]   David Moss, When All Else Fails: Government as the Ultimate Risk Manager 50 (2004).

[17]   Anastasia Telesetsky, *Insurance as a Mitigation Mechanism: Managing International Greenhouse Gas Emissions through Nationwide Mandatory Climate Change Catastrophe Insurance*, 27 Pace Environmental Law Review 691 (2010).

[18]   Howard Kunreuther, *Mitigating Disaster Losses through Insurance*, 12 Journal of Risk and Uncertainty 171 (1996).

Mortgage Regulations in 1998, stating that before a mortgage contract was concluded, the mortgagor was required to obtain household insurance or to relegate this task to the mortgagee (Article 25). However, in 2006, following lobbying from the banks to expand their residential mortgage businesses, the China Banking Regulatory Commission issued a notice relaxing the requirement that banks must stipulate that residential mortgage insurance is mandatory.[19] In view of that history, the government should not require banks to impose catastrophe coverage as a condition of credit but rather impose a disaster insurance requirement directly on property owners.

Mandatory catastrophe insurance would help solve the general lack of insurance take-up in China. In this regard, it is instructive to compare the experiences of California and Chile. In California, there is no requirement that residents in seismic areas should purchase earthquake insurance, and banks do not require it as a condition for house mortgages. As of 2015, only 10.23 percent of Californians living in earthquake-prone areas have residential earthquake policies.[20] Chile's practices, by contrast, have been far more successful. In 2010, Chile's take-up rate for residential mortgage holders—about a quarter of all of Chile's four million homes— was 96 percent because they are strongly advised to purchase earthquake insurance.[21] Meanwhile, among homeowners *without* a mortgage, only 3 percent were insured to some extent against earthquakes.[22]

This comparison suggests that requiring catastrophe coverage would vastly increase the percentage of Chinese property owners who are insured. To make such a mandate more palatable to citizens, it would be further advisable to bundle flood and earthquake insurance into ordinary homeowners' coverage. Together, these two measures would dramatically increase take-up rates.[23]

---

[19] Johanna Hjalmarsson and Mateusz Bek, *Legislative and Regulatory Methodology and Approach: Developing Catastrophe Insurance in China* in Insurance Law in China 202 (Johanna Hjalmarsson and Dingjing Huang eds., 2015).

[20] D. Marshall, *An Overview of the California Earthquake Authority*, 21 Risk Management and Insurance Review 73 (2018).

[21] Michael Useem, Howard Kunreuther and Erwann Michel-Kerjan, Leadership Dispatches: Chile's Extraordinary Comeback from Disaster 121–122 (Stanford University Press, 12015).

[22] Michael Useem, Howard Kunreuther and Erwann Michel-Kerjan, Leadership Dispatches: Chile's Extraordinary Comeback from Disaster 121–122 (Stanford University Press, 2015).

[23] For example, see H. Kunreuther, *All-Hazards Homeowners Insurance: Challenges and Opportunities*, 21 Risk Management and Insurance Review 141 (2018).

B.   Public–Private Catastrophe Insurance Partnerships

Insurers are in a better position than the government to provide incentives for disaster risk mitigation and thereby reduce costs in compensating victims. However, insurers are reluctant to provide coverage for catastrophic correlated risks due to the inherent challenges of climate disasters. Catastrophe insurers face both supply and demand barriers to developing new products and to providing adequate coverage. This justifies intervention by, or partnership with, the Chinese government to support disaster insurance. A public–private partnership where government intervention contributes to solving both supply and demand barriers of insurance development would be an optimal choice in practice.

1.   Types of public–private partnerships in catastrophe insurance

There is no, one definition of the term "public–private partnership", and the meaning of this term has shifted from being a simple technique adopted by the government, to being a comprehensive policy preference at the heart of governance.[24] In practice, there are different types of public–private partnership in the insurance realm. It is worth examining these different models and to draw lessons for China.

In the first form of public-private partnership, insurers provide disaster risk coverage and compensate victims in the case of damages, while the government does not intervene in either direct insurance or reinsurance. What the government undertakes is the adoption of administrative measures to facilitate and guarantee the independence of insurers' operations. The old English flood insurance program, that is, the one that existed before the entry of the new Flood Re Program in the United Kingdom, is a typical model. The UK's private flood insurance scheme demonstrates how a largely private insurance scheme could work, and it has proven to be efficient.[25] The challenge facing the United Kingdom, however, is how to keep it widely available and affordable, specifically to enable high-flood-risk households to obtain such insurance at an affordable price, which is a main focus of the new Flood Re Program.[26]

---

[24]   Anthony E. Boardman, Carsten Greve and Graeme A. Hodge, *Comparative Analyses of Infrastructure Public–Private Partnerships*, 17 Journal of Comparative Policy Analysis: Research and Practice 441 (2015).

[25]   Michael Huber, *Insurability and Regulatory Reform: Is the English Flood Insurance Regime Able to Adapt to Climate Change?*, 29 Geneva Papers on Risk and Insurance—Issues and Practice 169 (2004).

[26]   The Flood Re model is loosely based on Pool Re, a reinsurance scheme for terrorism risks formed in 1993 in response to the threat posed by the Irish

In the second form of public–private partnership, the public sector plays a greater role than in the first form. Private insurers underwrite disaster risk like many other lines in the private market, while the government acts as the last resort to provide additional capacity through reinsurance or other type of financial guarantee. The French Caisse Centrale de Réassurance (CCR), US federally-backed terrorism insurance, and the Turkish Catastrophe Insurance Pool (TCIP) are typical models. Under the CCR, French insurers are responsible for underwriting primary coverage, while the government provides subsidized reinsurance with an unlimited guarantee and cooperates with private insurers to create prevention and mitigation plans.[27] This enables primary insurers to underwrite disaster insurance policies at affordable prices for homeowners.[28] The challenge of this form of insurance is that government-sponsored reinsurance generally offers subsidized premiums, which reduces some of the risk mitigation incentives for primary insurers.[29]

In the third form of public–private partnership, government becomes the actual and major player in the catastrophe insurance program. The government is the primary risk bearer for disasters, while private insurers only play an exclusively administrative role in running the program.[30] This form is exemplified by the National Flood Insurance Program (NFIP) in the United States. The NFIP was established according to the National Flood Insurance Act of 1968, in order to assume flood risk and offer coverage for disaster-prone-area residents.[31] Private insurers do not assume risks but only administer policy coverage as the agents of the Federal Emergency Management Agency (FEMA). Moreover, there is no reinsurance arrangement in the NFIP. If claims exceed its financial capacity, the federal government provides

---

Republican Army and other terrorist activity. See Johanna Hjalmarsson and Mateusz Bek, *Legislative and Regulatory Methodology and Approach: Developing Catastrophe Insurance in China* in Insurance Law in China 197 (Johanna Hjalmarsson and Dingjing Huang eds., 2015).

[27] Act No. 82-600 of 13 July 1982 on the Indemnification of Victims of Natural Catastrophes, JORF 14 July 1982, 2242.

[28] Erwann Michel-Kerjan, *Catastrophe Economics: The National Flood Insurance Program*, 24 Journal of Economic Perspectives 165 (2010).

[29] Peter Molk, *The Government's Role in Climate Change Insurance*, 43 Boston College Environmental Affairs Law Review 411, 424–425 (2016).

[30] Peter Molk, *The Government's Role in Climate Change Insurance*, 43 Boston College Environmental Affairs Law Review 411, 424 (2016).

[31] Howard Kunreuther, *The Role of Insurance in Reducing Losses from Extreme Events: The Need for Public–Private Partnerships*, 40 Geneva Papers on Insurance 741 (2015).

a bailout.[32] Compared to China's Whole-Nation System—a pure, government-provided compensation program—the US approach provides only a modest advantage since private insurers have no incentives to regulate policyholders' behavior and so enhance the effectiveness of the program.

2.   Guiding principles for public–private partnerships in China

In order to establish a feasible and sustainable public–private catastrophe insurance partnerships in China, the following two principles are relevant for utilizing public–private partnerships as a risk-bearing tool that encourages policyholders to invest in cost-effective mitigation, while at the same time solving issues of affordability.

Principle 1. Insurers as private risk regulators   Given that catastrophe losses are (at least partially) linked to human activities, such as migration to vulnerable areas where problems of aging infrastructure exist and there is insufficient investment in risk-reduction measures, insurance has the potential to control risk and the associated losses by deterring risky behaviors. If moral hazard could not be controlled, liability insurance for major GHG emitters, for example, could not compel timely climate change mitigation. At the same time, this may have socially negative consequences as it could dilute the incentives of the emitters (the potential tortfeasors for climate disaster victims) to invest in prevention.[33] Property/casualty insurance would have the same effects.

Regulation by insurance is essential to align the interests of both parties in disaster risk reduction. Fortunately, insurers have the incentives and capacity to minimize payouts to the insured in order to maximize their profits: "If an insurer can lower its premiums by lowering its risk of paying claims, it can undercut its competitors by charging lower premiums, thereby attracting more business. Marketplace considerations, rather than altruism, drive insurers to reduce risk."[34] For example, through insurance rate classification, liability insurers can charge experience-rated

---

[32]   Peter Molk, *The Government's Role in Climate Change Insurance*, 43 Boston College Environmental Affairs Law Review 411, 424–425 (2016).

[33]   Anastasia Telesetsky, *Insurance as a Mitigation Mechanism: Managing International Greenhouse Gas Emissions through Nationwide Mandatory Climate Change Catastrophe Insurance*, 27 Pace Environmental Law Review 691 (2010); Steven Shavell, *On Liability and Insurance*, 13 Bell Journal of Economics 120 (1982).

[34]   John Aloysius Cogan Jr, *The Uneasy Case for Food Safety Liability Insurance*, 81 Brooklyn Law Review 1495, 1522 (2016).

premiums and thus induce policyholders to behave more carefully than they would otherwise. In practice, insurance laws' reluctance to prohibit rate classification based on controllable characteristics supports insurers' behavior-control functions. Through the payment of an insurance premium before a disaster, the insured is made aware of its vulnerability to the disaster. This has a positive impact on insureds' behavior.[35] Compared to the state, insurers have the capacity to manage moral hazard because of both superior information and competition.[36]

Meanwhile, insurers have many techniques, such as risk-based pricing, deductibles, exclusions, and loss-reduction services, to reduce risk, and the consensus is that these techniques work reasonably well.[37] Theoretically, risk-based pricing is regarded as the central approach of insurers to risk mitigation. Risk-based pricing can provide individuals with financial incentives as to the degree of the hazards they face and encourage them to engage in cost-effective mitigation measures to reduce their vulnerability. According to insurance actuarial studies, an insurer's price-setting process gives policyholders accurate signals. Since the insurance premium is based on expected overall losses, derived by multiplying loss probability by loss severity, reducing either the probability or the severity of loss may lower the premium.[38] As long as such reduction cost is lower than the discount of the premium, policyholders would likely undertake mitigation measures to reduce loss probability and/or severity.[39] If the loss probability and

---

[35] George Priest, *The Government, the Market and the Problem of Catastrophic Loss*, 12 Journal of Risk and Uncertainty 221 (1996) (arguing that private insurance is able, via the control of moral hazard by insurers, to provide incentives for mitigation of disaster risks).

[36] By utilizing the methodologies of actuarialism, private contracting, and ex post claim investigation, insurers can easily collect customers' purchasing information, thereby replacing the government in this role. See Omri Ben-Shahar and Kyle D. Logue, *Outsourcing Regulation: How Insurance Reduces Moral Hazard*, 111 Michigan Law Review 197 (2012). It is generally believed that insurance markets tend to be highly competitive with respect to price. See Daniel Schwarcz, *Regulating Consumer Demand in Insurance Markets*, 3 Erasmus Law Review 23, 43 (2010).

[37] Tom Baker and Peter Siegelman, *The Law & Economics of Liability Insurance: A Theoretical and Empirical Review* in Research Handbook in the Economics of Tort 169 (Jennifer Arlen ed., 2013).

[38] Peter Molk, *Private Versus Public Insurance for Natural Hazards: Individual Behavior's Role in Loss Mitigation* in Risk Analysis of Natural Hazards 265, 265–277 (Paolo Gardoni et al. eds., 2016).

[39] Peter Molk, *Private Versus Public Insurance for Natural Hazards: Individual Behavior's Role in Loss Mitigation*, in Risk Analysis of Natural Hazards 265, 265–277 (Paolo Gardoni et al. eds., 2016).

loss severity are too high, however, insurers may refuse to underwrite in the first place.[40]

Through offering effective incentives, applying regulatory techniques, and monitoring policyholders' behavior, insurers have the capacity to remedy the risk of moral hazard, promote policyholders' cost-effective actions and thus contribute to climate disaster mitigation.

Principle 2. The government as last resort   Steps need to be taken by the government to enable insurers to provide disaster coverage to those residing in hazard-prone areas and to help poorer people to pay for that insurance. Meanwhile, the government may also have a play to role in providing protection against catastrophic losses that cannot be covered by private catastrophe insurers.

The Chinese government could act as reinsurer of last resort by helping fill the "capacity gap" of primary insurers in underwriting climate disasters. Compared to the private sector, the government has "a deep credit capacity" and can raise money effectively and quickly through tax or by issuing debt or government bonds following disasters.[41] It is widely recognized that the government has the capacity to avoid private reinsurance's underwriting cycles and thereby promote primary insurers' supply of catastrophe coverage.[42] Meanwhile, the government as reinsurer would not interfere with a private insurer's business, including underwriting policies, collecting premiums, and paying claims, which will ensure that the private insurance market can continue to play a leading role in climate disaster adaption and mitigation.[43] In practice, a model of reinsurance by government is developed in many countries to deal with earthquake

---

[40]   This also serves as a gatekeeping function. See Tom Baker and Thomas O. Farrish, *Liability Insurance & the Regulation of Firearms*, in Suing the Gun Industry: A Battle at the Crossroads of Gun Control and Mass Torts 292, 294–295 (Timothy D. Lytton ed., 2008).

[41]   Louis Kaplow, *Incentives and Government Relief for Risk*, 4 Journal of Risk and Uncertainty 167 (1991).

[42]   Reinsurance's underwriting cycles refers to the tendency of insurance markets to go through alternating phases of "hard" and "soft" markets. A hard market leads to decreased supply but increased premium, whereas in a soft market, coverage supply is plentiful and prices decline. See David Cummins and Olivier Mahul, Catastrophe Risk Financing in Developing Countries: Principles for Public Intervention 194 (2009).

[43]   Michael Faure, *Climate Change Adaption and Compensation* in Research Handbook on Climate Change Adaptation Law 134 (Jonathan Verschuuren ed., 2013).

and terrorism risks.[44] It is reasonable to assume that it could be a feasible method to deal with climate disaster risks in China as well.

Meanwhile, the government should act as last resort in helping low-income to pay for insurance. Only when insurers charge risk-based premiums will policyholders have financial incentives to take adaptation and mitigation measures to reduce premiums.[45] However, this poses challenges to low-income residents who have to live in vulnerable areas due to budget constraints.[46] The government could provide insurance vouchers to those residents, which could cover part of the cost of insurance, and at the same time maintain insurance premiums reflecting the degree of risk.[47]

C. Modifying the Shenzhen Model[48]

In 2014, Shenzhen began its large-scale experiment with a nationally funded city-wide model called the Disaster Insurance Pilot, which is the first experiment by the Chinese government to spread risks associated with climate change events through catastrophe insurance at the municipal level.[49] Shenzhen was selected for the pilot because its geographical location means that it is vulnerable to a variety of potential disaster threats, a large number of valuable assets, and is experimenting with a carbon market to reduce greenhouse gas emissions.[50]

---

[44] For example, Turkey (Turkish Catastrophe Insurance Pool), Japan (Japan Earthquake Reinsurance Co), United Kingdom (Flood Re, where insurers can cede the riskiest properties to the Flood Re pool at a discounted price), United States (Terrorism Risk Insurance Program and the Florida Hurricane Catastrophe Fund, examples of state-level reinsurance).

[45] Howard Kunreuther, *The Role of Insurance in Reducing Losses from Extreme Events: The Need for Public–Private Partnerships*, 40 Geneva Risk and Insurance Review 741, 758 (2015).

[46] Howard Kunreuther, *The Role of Insurance in Reducing Losses from Extreme Events: The Need for Public–Private Partnerships*, 40 Geneva Risk and Insurance Review 741, 758 (2015).

[47] Howard Kunreuther and Rosemary Lyster, *The Role of Public and Private Insurance in Reducing Losses from Extreme Weather Events and Disasters*, 19 Asia Pacific Journal of Environmental Law 29, 42–43 (2016).

[48] The description of the Shenzhen Model draws on the co-author's article: see Anastasia Telesetsky and Qihao He, *Climate Change Insurance and Disasters: Is the Shenzhen Social Insurance Program a Model for Adaptation?*, 43 Boston College Environmental Affairs Law Review 485 (2016).

[49] Susan Munro and Amy Wang, *China Announces Policies to Accelerate Development of Chinese Insurance Industry*, Steptoe & Son (July 15, 2014), available at http://www.steptoe.com/publications-9727.html.

[50] Jiajun Wang, *Research on Legislation Pattern of Catastrophe Insurance in Major City of China—Based on Study on Pilot Work of Catastrophe Insurance in*

This catastrophe insurance framework includes three different inter-acting tiers of insurance. The first tier is the government's catastrophe assistance insurance, which is bought by the Shenzhen municipal government from the People's Insurance Company of China (PICC) Shenzhen branch to supply basic assistance to all residents. This insurance functions as indemnity insurance: the insurance company will pay for actual medical costs and pension costs due to victims' injuries or deaths caused by the disasters. The second tier is a catastrophe fund managed by the city of Shenzhen. Capital sources of the catastrophe fund include government appropriations, social donations, and investment earnings. The third and final tier is pure private catastrophe insurance to cover property losses, in contrast to the first and second tiers of the Shenzhen insurance framework, which constitute government-funded policy coverage.

The Shenzhen Model is a good start, but is not yet perfect. It provides a target of opportunity to implement a multi-layered public–private catastrophe insurance partnership for reducing climate disaster risks. The proposed New Shenzhen Model according to its two Guiding Principles is intended to provide a foundation for China's future catastrophe insurance in light of the interests that private insurers have in providing insurance coverage to residents.

The New Shenzhen Model should be run by private insurers rather than the government in order to promote risk mitigation. Under this multi-layered public–private catastrophe insurance partnership, the tiers of risk transfer need to be supported by public and private sector activity centered on risk communication and risk reduction.[51]

Under this modified Shenzhen approach, the first tier of catastrophe losses would be covered by insureds' self-insurance (in the form of co-pays and deductibles). The first tier would provide mitigation incentives and prevent moral hazard.

The second tier would consist of mandatory catastrophe insurance provided by private insurers charging risk-based premiums. Putting the private insurance sector in the second-loss position would result in more accurate pricing of catastrophe risk and activate regulation of climate change risk by insurance while increasing loss mitigation.

The third tier would be private (both international and domestic) reinsurance and insurance-linked securities. These two instruments would

---

Shenzhen, 29 Journal of the Postgraduate of Zhongnan University of Economics and Law 81, 81–87 (2013).

[51]   Carolyn Kousky and Howard Kunreuther, *Risk Management Roles of the Public and Private Sector*, 21 Risk Management and Insurance Review 181 (2018).

help spread the cost of catastrophe losses covered by primary insurers in the private market and provide substantially greater capacity.

Finally, the fourth tier would be a government-funded backstop (i.e., government-sponsored reinsurance), which would assume the risks of any losses above a predetermined threshold, thereby capping private insurers' maximum exposure. This type of cap could promote the initial entry of private insurers and also act as a last resort.[52]

V. CONCLUSION

This is the conclusion of both this chapter and the book. I have weaved the entire analysis into a roadmap for reform and proposed policy recommendations for China. Now, I would like to review the big picture of the book: that command-and-control regulation and disaster relief by the government is not enough to respond to climate change and that private insurance offers a promising role both in compensation and in loss mitigation.

Government is playing an expanding role in catastrophe disaster aid, relief, and compensation around the world. The money paid out by government following disasters is rising sharply, and this trend is evident in the United States, European countries, and many others.[53] The same is happening in China, where government has traditionally played a fundamental role in dealing with catastrophe disasters. Nonetheless, the government approach using the Whole-Nation System to manage catastrophe disasters needs reform. The question of whether insurance works better than the government in managing catastrophe risks in China and can therefore be recommended to policyholders has to be answered in the positive. It might be appropriate to combine the Whole-Nation System with catastrophe insurance to marry the merits of both private market and

---

[52] "The experience of both Flood Re and TRIA suggests that if the government can cover the riskiest properties, or cover catastrophic losses, the private market could profitably offer disaster insurance for the remaining losses above the deductible". See Carolyn Kousky and Howard Kunreuther, *Risk Management Roles of the Public and Private Sector*, 21 Risk Management and Insurance Review 181 (2018).

[53] David Cummins, M. Suher and G. Zanjani, *Federal Financial Exposure to Natural Catastrophe Risk* in Measuring and Managing Federal Financial Risk 61 (D. Lucas ed., 2010); European Commission, *Disaster Risk Reduction: Increasing resilience by reducing disaster risk in humanitarian action*, DG ECHO Thematic Policy Documents (2013), available at http://ec.europa.eu/echo/files/policies/prevention_preparedness/DRR_thematic_policy_doc.pdf.

public government. It seems clear that the Whole-Nation System mainly works well in emergency relief. Beyond that, the government should encourage catastrophe insurance to be responsible for risk finance and loss compensation and support insurance in that role.

Insurance is one of mankind's greatest inventions, an extraordinarily useful tool to reduce risk.[54] Insurance is a good device to transfer risk for individuals, who can substitute a small certain cost, the premium, for a large uncertain financial loss. Compared with government intervention, insurance is better equipped to deal with catastrophe risks due to its advantages of lower transaction costs, lower adverse selection, and more efficiency as a result of competitive markets. It is generally agreed that insurance is a good compensation model that provides victims with financial protection in catastrophe disasters.

Meanwhile, insurers increasingly act as private risk regulators, substituting or complementing public regulation. Five technical tools have been identified in the previous chapters that can be employed by insurers on the one hand to control the moral hazard risk, and on the other hand to provide incentives for disaster risk reduction (risk-based pricing, contract design, loss prevention, claims management, and refusal to insure). In line with the literature claiming that insurers act as private regulators,[55] I argued that when these technical tools are indeed effectively applied, insurers can fulfil their task in contributing to disaster risk reduction. However, when I then examined the possibilities in specific countries (the United Kingdom, France, the United States, Japan, and Turkey) to apply these technical tools, I noticed that the possibilities to do so in practice are often limited, precisely as a result of government intervention. Government intervention would, for example, prohibit premium differentiation (to promote affordability of insurance) or prohibit a refusal to insure (in order to guarantee an equal access to catastrophe insurance for all citizens). As a result of these restrictions following from government intervention, insurers in many legal systems often cannot fully play their role as private risk regulators.

It is of course too early simply to conclude that government interventions are necessarily undesirable. However, the interesting conclusion is that it is possible to combine the political desiderata (for example, of providing affordable disaster insurance to all) in a model whereby insurers

---

[54]   Howard Kunreuther, Mark V. Pauly and Stacey McMorrow, Insurance and Behavioral Economics: Improving Decisions in the Most Misunderstood Industry 13 (2013).

[55]   Omri Ben-Shahar and Kyle D. Logue, *Outsourcing Regulation: How Insurance Reduces Moral Hazard*, 111 Michigan Law Review 197 (2012).

could still apply their technical tools aimed at disaster risk reduction. Government acting as last resort is a reasonable choice since not only can it support failing catastrophe insurance due to its deep credit capacity (as the last resort) but it can also regulate primary insurers' behaviors in risk mitigation and risk management through reinsurers' regulatory activities. More importantly, that would allow insurers to play an important role as private regulators, thus substituting or complementing public regulation aimed at disaster risk reduction. This private–public partnership becomes a prototype to develop catastrophe insurance in many countries. It should be developed in China as soon as possible to cope with the increasing catastrophe risks.

Index